DATE DUE

MAY 0 1 2003		
MAY 0 1 2003		
APR 2 8 2006		
APR 2 3 2010		
JUN 3 0 2010		

Environmental Assessments and Statements

Environmental Assessments and Statements

John E. Heer, Jr.

Professor and Chairman, Civil Engineering
University of Louisville

D. Joseph Hagerty

Associate Professor of Civil Engineering
University of Louisville

VAN NOSTRAND REINHOLD ENVIRONMENTAL ENGINEERING SERIES

 VAN NOSTRAND REINHOLD COMPANY
New York   Cincinnati   Atlanta   Dallas   San Francisco
London   Toronto   Melbourne

Van Nostrand Reinhold Company Regional Offices:
New York   Cincinnati   Atlanta   Dallas   San Francisco

Van Nostrand Reinhold Company International Offices:
London   Toronto   Melbourne

Library of Congress Catalog Card Number: 76-30603
ISBN: 0-442-23030-3

Manufactured in the United States of America

Published by Van Nostrand Reinhold Company
450 West 33rd Street, New York, N.Y. 10001

Published simultaneously in Canada by Van Nostrand Reinhold Ltd.

15 14 13 12 11 10 9 8 7 6 5 4 3 2 1

Library of Congress Cataloging in Publication Data

Heer, John E
  Environmental assessments and statements.

  (Van Nostrand Reinhold environmental engineering
series)
  1. Environmental law—United States.   2. Environmental protection—United States.   3. Environmental
impact statements.   I. Hagerty, D. J., joint author.
II. Title.
KF3775.H4   344'.73'046   76-30603
ISBN 0-442-23030-3

Dedicated
To our wives
Agnita and Patricia

Van Nostrand Reinhold Environmental Engineering Series

THE VAN NOSTRAND REINHOLD ENVIRONMENTAL ENGI-
NEERING SERIES is dedicated to the presentation of current and vital
information relative to the engineering aspects of controlling man's physical environment. Systems and subsystems available to exercise control of
both the indoor and outdoor environment continue to become more so-
phisticated and to involve a number of engineering disciplines. The aim of
the series is to provide books which, though often concerned with the life
cycle—design, installation, and operation and maintenance—of a specific
system or subsystem, are complementary when viewed in their relationship to the total environment.

Books in the Van Nostrand Reinhold Environmental Engineering
Series includes ones concerned with the engineering of mechanical
systems designed (1) to control the environment within structures, including those in which manufacturing processes are carried out, (2) to
control the exterior environment through control of waste products ex-
pelled by inhabitants of structures and from manufacturing processes.
The series will include books on heating, air conditioning and ventilation,
control of air and water pollution, control of the acoustic environment,
sanitary engineering and waste disposal, illumination, and piping systems
for transporting media of all kinds.

Van Nostrand Reinhold Environmental Engineering Series

ADVANCED WASTEWATER TREATMENT, by Russell L. Culp and Gordon L. Culp

ARCHITECTURAL INTERIOR SYSTEMS—Lighting, Air Conditioning, Acoustics, John E. Flynn and Arthur W. Segil

SOLID WASTE MANAGEMENT, by D. Joseph Hagerty, Joseph L. Pavoni and John E. Heer, Jr.

THERMAL INSULATION, by John F. Malloy

AIR POLLUTION AND INDUSTRY, edited by Richard D. Ross

INDUSTRIAL WASTE DISPOSAL, edited by Richard D. Ross

MICROBIAL CONTAMINATION CONTROL FACILITIES, by Robert S. Runkle and G. Briggs Phillips

SOUND, NOISE, AND VIBRATION CONTROL, by Lyle F. Yerges

NEW CONCEPTS IN WATER PURIFICATION, by Gordon L. Culp and Russell L. Culp

HANDBOOK OF SOLID WASTE DISPOSAL: MATERIALS AND ENERGY RECOVERY, by Joseph L. Pavoni, John E. Heer, Jr., and D. Joseph Hagerty

ENVIRONMENTAL ASSESSMENTS AND STATEMENTS, by John E. Heer, Jr. and D. Joseph Hagerty

ENVIRONMENTAL IMPACT ANALYSIS: A New Dimension in Decision Making, by R. K. Jain, L. V. Urban and G. S. Stacey

CONTROL SYSTEMS FOR HEATING, VENTILATING AND AIR CONDITIONING (Second Edition), by Roger W. Haines

Acknowledgments

In the course of preparing a book such as this many individuals lend their assistance, advice, and support. The preparation of this book was no exception and the authors are grateful and deeply indebted to all who assisted them in any way. To acknowledge everyone who contributed to this manuscript would be an impossible task; however, the assistance and generosity of several individuals were of such magnitude that they must be mentioned.

Mr. Frank Christ and Mr. William Leegan of the Louisville District Office, U.S. Army Corps of Engineers and Mr. Paul H. Leskinen of the Greenfield Village and Henry Ford Museum, Dearborn, Michigan, assisted the authors in obtaining the photographs for the figures.

Special and sincere thanks must also go to Ms. LeAnne Whitney, Mrs. Betty Weber, and Mrs. Donna Greenwell, who spent many hours typing, retyping, and proofreading the manuscript. Without the assistance of friends such as these, this book would not have been possible.

JOHN E. HEER, JR.
D. JOSEPH HAGERTY

viii

Preface

All human activity involves interaction between man and his physical surroundings. Energy must be consumed and materials must be utilized to sustain life. Competition with other life forms is inevitable. Man must create impacts on living and nonliving sectors of his environment in order to survive.

This book has a dual purpose. First, it has been our purpose to describe ways and means by which the environmental effects of specific human activities may be assessed and evaluated. Second, we have attempted to show that such assessment and evaluation must not be an end in itself. When identified, environmental impacts must be ameliorated as much as possible. Resources must be conserved and recovered; degradation of natural systems must be minimized, if not eliminated. Plans may require revision. Occasionally, designs may require revision or rejection.

It is our hope that this book will assist concerned individuals in identifying environmental impacts, systematically evaluating them, and developing practicable solutions to identified problems. We believe that such solutions can and will be achieved.

J. E. HEER, JR.
D. J. HAGERTY

Contents

Environmental Assessments and Statements

1

Introduction

1-1. GENESIS

The first chapter of this text is devoted to a description of the origins, purposes, and limitations of this text, and to the establishment of the historical perspective for assessment of environmental impact. Above all, the necessity for such impact assessment is presented.

1-1.1 Definition of Assessment

This book is devoted to a description of the purposes, means, and methods for assessment of environmental impacts of contemplated actions of man. It is appropriate, then, in the beginning to define the process of assessment. *Webster's New Twentieth Century Dictionary of the English Language* defines the act of assessment as "an evaluation." Further, to assess is to ". . . fix the value of. . . ." In essence, an assessment is a quantitative evaluation for the purpose of establishing relative values among various quantities or parameters. In environmental assessment, parameters are chosen to reflect the quality of the environment and the effects, favorable or unfavorable, of proposed actions on that quality. Thus, environmental assessment consists in establishing quantitative values for selected parameters which indicate the quality of the environment before, during, and after the action.

With a definition of assessment, as such, comes an almost automatic question, "Why assess?" At the present time, in the United States and in the world in general, considerable concern is being felt for the end consequences of all human activities. Some individuals now are questioning growth itself and are seeking to establish the limits to growth. Other persons are concerned not with limits, but with the directions in which that growth is taking place. With an awakening of concern and interest in the possible consequences of man's actions on his habitat, and on his chances even for survival, a mechanism is being sought to answer these questions concerning the directions and limitations to growth which seem to be plaguing so many individuals today. As in any human endeavor, the acquisition of knowledge concerning the variables in a difficult situation constitutes the first step in the solution of the problem. Such acquisition of knowledge often can be achieved through the identification and quantification of crucial parameters or variables; a quantity is essentially an assessment. However, this simple statement is inadequate; even in the assessment process itself, a question arises as to the limits of activity to be undertaken. In other words, in identifying and quantifying those parameters which are crucial to establishing the quality of life in a given environment, to what degree must the identification and evaluation process be pursued?

1-1.2   Extent of Assessment

In assessing the effects of a particular activity on the environment, and in assessing the quality of a given environment, several questions arise concerning the degree and complexity of the assessment operation. In evaluating parameters which characterize the quality of the environment, the investigator must decide which parameters are significant and which are insignificant. Moreover, he must decide when changes in the chosen parameters are significant or insignificant. With respect to the degree of investigation, the direct effects of a proposed action must be assessed, but, additionally, the indirect or secondary effects of a contemplated action must also be examined in any comprehensive assessment. Furthermore, in some cases, even third-order or tertiary effects may be deemed significant and worthy of investigation. The limit to the degree of investigation (third-order, fourth-order, etc.) must be decided on the basis of yet another assessment. Are the secondary, tertiary, etc., effects significant or insignificant? Obviously, in situations of this kind, value judgments must be made. Because the very nature of an assessment process includes many successive value judgments, it is inappropriate to attempt to fix a standard procedure for all such assessments. Therefore, in this book, no such attempt is made. Rather, efforts are made to present the basic concepts associated with comprehensive and correct evaluation and assessment techniques. It is the intention of the authors to present the philosophy and concepts of environmental assessment, rather than to describe a "cookbook" procedure for environmental assessment.

1-1.3 Purpose of Assessment

The emphasis in this book is placed upon the philosophy and concepts of assessment rather than on standardized procedures and formulas for evaluation. Thus, a major topic must be the overall purpose of environmental assessment itself. The basic reason for such assessment is to allow an informed public and professional to make intelligent choices among alternative courses of action. That is, an environmental assessment is intended to present a clear and concise picture of the benefits and costs in terms of natural and cultural assets, as well as the social values associated with various alternative courses of action. Additionally, as utilized in planning processes in the United States, environmental assessments have come to be a valuable tool in eliminating or mitigating undesirable effects on the environment that arise from contemplated actions. In many cases, a thorough and comprehensive environmental assessment has led to the identification of features in proposed projects which would produce severely detrimental effects on the surrounding environment. Because of the early identification of such effects, modifications in the design and construction of the proposed facility have been possible and the detrimental consequences have been lessened or eliminated. Additionally, through the performance of an environmental assessment, it has been possible to identify various operational modes for proposed facilities. In this sense, the various operational modes constitute alternatives to a proposed action and therefore should be fully evaluated for possible benefits and costs. Finally, the performance of a comprehensive environmental assessment will lead to a very definite benefit in identifying the consequences of maintaining the status quo. It is a truism that an often neglected alternative in the planning of human activity is the so-called null alternative; i.e., the course of action consisting of no action. In many instances, it has been found that a continuance of present life style, resource utilization, or similar activities is preferable to a wide range of proposed alternative actions which would change the status quo. This identification of the null alternative and its possible beneficial effects has been one of the most valuable facets of the environmental assessment process as developed to date in the United States.

1-1.4 Origins, Scope and Purposes of This Book

This book has grown out of the experiences of the authors in preparing environmental impact assessments for various private, local, state, and federal agencies. In these assessment efforts, the authors consistently have found several deficiencies. There is a striking lack of standards for the proper evaluation of environmental quality, both from the point of view of crucial parameters and of limiting values for the identified crucial parameters. Additionally, there is a lack of coherent methodology for logical and comprehensive environmental assessment. Finally, there is a striking lack of coordination among the various segments of most environmental assessment teams, between team members and personnel of agencies initiating the action under investigation, and between the agency personnel and the general public. These deficiencies, in part, stem from a general ignorance of the value of environmental assessment on the part of the lay public. Additionally, some of these deficiencies stem from a lack of knowledge, on the part of initiating agency personnel and assessment specialists, concerning all of the provisions of current environmental legislation and environmental technology. There is a common ignorance of the potential for changes in the design and operation of proposed facilities which can result from a properly performed initial impact assessment. In addition, there is the hostility, bred from lack of trust, which has developed between environmental groups and personnel from the agencies that initiate actions deemed to have significant effects on the environment. On the one hand, agency bureaucrats develop defensive postures at the first onslaught of criticism of their proposed activity, and, on the other hand, ill-informed so-called environmentalists frequently mount their white horses and charge forth in a quixotic crusade to "save the environment." Because of these factors (overall deficiencies and ignorance of proper procedures), there is an absolute necessity for competent and expert assessment of the environmental effects of contemplated activities.

Because of the necessity of the assessment mentioned above, this book has been compiled to provide the reader with a general acquaintance with environmental legislation and with the National Environmental Policy Act of 1969, in particular. This acquaintance with environmental legislation will set a framework for an appreciation of the maze of legal attempts at environmental protection which have been promulgated by local and national lawmakers and which, in many cases, have been enforced by court actions. It is essential that a thorough knowledge of environmental legislation be obtained by all members of an assessment team, by members of initiating agency groups, and by members of the public, since all of these individuals are intimately involved either in the assessment procedure, or in the review of prepared assessments, or in the completion of the proposed action.

In addition to a general commentary on the evolution of environmental law in the United States to date, specific information is given herein, concerning the origin, purpose, and scope of the National Environmental Policy Act of 1969 (NEPA), since this Act and its requirement for the preparation of environmental impact statements forms the most significant piece of environmental legislation enacted to date in the United States. In addition to information on the various facets of NEPA, a considerable portion of this book is devoted to a discussion of methods for the conduct of the actual assessment procedure. Various established methodologies are reviewed and analyzed. The advantages and disadvantages of each established method are listed and the limitations of all established methodologies are described. Finally, recommendations for methodologies to follow in the preparation of environmental impact assessments are presented. In essence, however, the authors have found that there is no all-inclusive method suitable for the preparation of impact assessments for all types of projects in all areas of this country. There is no unfailing, all-inclusive, golden rule for the preparation of environmental impact assessments. Therefore, the authors have presented the fundamentals of impact assessment and have attempted to illustrate the basic philosophy embodied in correct and comprehensive assessment techniques.

1-2. HISTORICAL PERSPECTIVE

A review of man's accumulated knowledge of his prehistoric and historic pasts shows numerous examples of situations where he has attempted to oppose natural processes, with varying degrees of success. In other cases, man, either through ignorance or through design, has attempted to work with nature. These attempts have also met with varying degrees of success. A short review of this catalog of experience will provide a valuable insight into the possibilities for improving man's lot through proper assessment of the consequences of his actions.

1-2.1 Ancient Man

A considerable body of knowledge has been assembled by archaeologists and anthropologists concerning the history of man on earth during those times when no written records were prepared or preserved. In this body of accumulated knowledge are striking examples of ecological adaptation by man, on the one hand, and on the other hand, elimination of human cultures through failure to adapt to natural conditions. A broad analogy may be made between modern man and those Paleolithic peoples who attempted to maintain their standard of life through pursuit of the bison and other large game forms roaming the earth in prehistoric times. The big game hunters developed a technology for the production of projectile points from various crystalline rocks such as flint and other varieties of microcrystalline quartz. The relative skill and success of the Paleolithic hunters has become apparent through the discovery of numerous sites where large accumulations of big game carcasses and remains have been found in association with large projectile points and other artifacts from the paleohunters. However, a serious question develops concerning the overall success of those hunters. From the accumulated archaeological evidence, it appears that the Paleolithic big game hunters were so successful that they may have eliminated many of the ruminants with which their survival was so intimately associated. Consequently, many of the big game hunting cultures and traditions probably vanished as a result of depletion of the animal herds upon which they preyed. On the other hand, there is considerable evidence to show a much greater long-term degree of success in environmental adaptation on the part of Paleolithic peoples who combined hunting activities with gathering of vegetable foods and the pursuit of small game. Many of these hunter-gatherer peoples have survived in remote areas of the world down to the present day. This contrast, in the survival of hunter and hunter-gatherers, presents a clear example of the over-kill possibilities associated with rampant use of an apparently successful technological innovation; in this case, the flint projectile point. Many analogies with current technological advances could be drawn at this point.

Among the many other lessons which could be learned through a study of ancient man is an appreciation of the great influence of environmental factors on social organization. A striking example of this effect of environmental constraints was described over 130 years ago by John L. Stephens in his description of a journey through the Yucatan peninsula in Mexico. The Yucatan peninsula was the home of the ancient Maya culture, a culture which developed a calendar which was more accurate than that of the Spanish Conquistadors who overran the Maya civilization and destroyed it. During his travels in Yucatan, Stephens found many wells and water storage facilities, called cenotes, which formerly had been used to supply water to the large Mayan cities. In one striking example, he describes the difficulties associated with descending hundreds of feet into the bowels of the earth to obtain a bucket of water which then had to be carried along an arduous path back up to the surface, a journey which altogether required more than an hour to complete. This well formerly had supplied the water needs for a city of over 100,000 people. The constant flow of water carriers to and from this small source was evidenced by the deep grooves worn in the stone of the cave-well site by the tread of countless bare feet over many, many years. The amount and degree of social organization involved in obtaining a constant water supply in this manner should be obvious. Thus, in this particular instance, the environmental constraint of scarce water located at considerable depth in cavernous rock was responsible for the creation of a significant part of the Maya social structure (Ref. 1-9). Numerous other instances could be cited to show the overall social effect of environmental constraints as evidenced in ancient cultures.

1-2.2   Historical Man

As with the case of human cultures which are known only from their artifacts and material remains, in particular those cultures which have left a written record of their successes and failures and in those cultures which are available for study today, success and failure can be seen in the adaptations of life to various natural environments. For example, there are numerous references in the Bible to the extensive forests of great cedar trees in Lebanon. Today, pitifully sparse and widely separated groves are all that remain of the once great forests. In a similar way, the forests which covered the slopes of the Dolomites in Yugoslavia and the surrounding Balkan states have also disappeared in man's search for fuel and raw materials for manufacture. At one time, the European continent consisted of a marshy forest-covered region, described by Tacitus as a region of endless, inpenetrable forests on the other side of the Alps from sunny Italy (Ref. 1-1). It is very difficult to imagine dense forest and marshland covering the central portions of the European continent as we know them today. Thus, through history, continuous depletion of forest reserves and other natural resources has occurred. In many instances, the effects of deforestation have been irreversible and extremely detrimental. This detrimental utilization of land can be seen at the present time in those areas of Southeast Asia where slash-and-burn agriculture still is practiced. A striking example is found among the Khmer people in Cambodia and the surrounding mountains. The continued existence of this culture is in doubt because of the gradual depletion of forest cover, and the leaching and laterization of exposed soils through intensive agriculture practiced on the cleared land. On the other hand, in some instances, even slash-and-burn agriculture has been practiced with a certain degree of success and adaptation to the prevailing ecology of a given region. For example, Marvin Harris (Ref. 1-4) has described the virtually unique culture of the hill peoples of New Guinea, where the entire agricultural activity of the people is geared to the raising of pigs and the periodic mass slaughter of those pigs. For example, the Maring peoples plant a certain type of tree and begin to raise herds of swine as the initial step in an agricultural cycle. The pigs thrive in the timbered upland inhabited by the Marings, and the tribal women expend considerable effort in caring for the pigs. At the end of a period of approximately 10 to 12 years, when the size of the pig herd has increased to the point where the pigs are competing with their masters for food resources, the ritual tree is uprooted, a massive slaughter of swine takes place, and a feast is held for several days, during which the Maring and their allies gorge themselves on pig flesh. After the ritual slaughter of the swine and a long series of other preparations, the Marings go to war with neighboring villages to attempt to wrest control of territory from their neighbors. After a decisive battle has been fought, the vanquished tribes retreat from their former village sites and go to live and seek shelter with any allies who will accept them. The victors in the struggle do not occupy the villages and fields of the vanquished, but they do their utmost to destroy those villages and fields. The victors expand their own settlements slightly, but only intrude into the edges of the territory previously occupied by the vanquished tribe. The site of the defeated village then reverts to jungle, and for a period of 10 to 12 years the forest cover is regenerated, prior to another onslaught. of tribal warfare. During this time, the previously vanquished tribe may begin to filter back into their former haunts, and should they be successful in the ensuing cycle of warfare, they will move back into their former home sites. At that time, they slash down the trees and bushes that have grown in their former fields and burn them, thus introducing necessary nitrogen into the soil as a nutrient, from the ashes of the burnt vegetation. Scientific studies have shown that a period of approximately 10 to 12 years is sufficient for regeneration of the cropland in the Maring region. Thus, the apparently senseless cycle of pig farming, pig slaughter, and subsequent warfare has a cogent explanation in light of the environmental constraint on agricultural productivity of the forestland in the Maring region.

The foregoing examples show instances wherein man has created, unconsciously or consciously, irreversible damage to the natural environment with dire consequences for his own survival. These examples have also shown some instances of symbiosis, wherein even primitive peoples have managed to adapt their mode of life to survival within environmental constraints. At times, this adaptation to constraints has even taken the form of activities to appease the spirits of the hills, trees, plants, and animals of the primitive tribal habitat. The unconscious wisdom of this mode of life may be extremely valuable to modern man.

1-2.3   Great Successes and Failures

As a further extension of the previous analysis, the growth and development of the three earliest major civilizations known to man can be examined. These civilizations were quite similar in many ways, and they developed in river valleys along the Nile, Indus, and Tigris-Euphrates river systems. All of these civilizations were based, in part, upon recurrent flooding of the rivers at certain times of the year, with consequent siltation of low-lying lands along the river banks. In great measure, the civilizations in Mesopotamia along the Tigris and Euphrates Rivers and in India along the Indus River, failed, while that along the Nile River survived through many millenia up to present times. It is interesting to note the reasons for the apparent failure of two of these great early civilizations and the apparent success up to present times of the third. In large measure, the failure of the Indus civilization has been attributed to a deforestation of the Indian subcontinent and an over-grazing of the grassland which developed after the forest cover had been removed (Ref. 1-3, Ref. 1-6). Additionally, some indication has been seen that gradual build-up of soluble salts along the river valley has decreased the natural fertility of the soil in the Indus Valley. A cycle has been identified in which early settlers used the spring flooding of the river for irrigation and for soil enrichment, developing an extensive agriculture. With expanding population, more land was cleared with subsequent loss of timber and creation of grassland and cropland. As mentioned previously, over-grazing in the grassland and more frequent flooding and eventual salinization of croplands led to decreases in fertility and a failure of the agricultural system to support the ever-increasing population. Archaeological investigations have shown a gradual decay in the Indus Valley civilizations with apparent great civil disorder in the latter stages of the decline.

In the Tigris-Euphrates Valleys, a somewhat similar situation existed, in that spring floods irrigated the land and deposited a silt cover over the inundated areas. Nomadic peoples came out of West Central Asia and moved into Mesopotamia to exploit the agricultural potential of the river valleys. Settlements sprang up along the river valleys, but these settlements, which eventually became cities, were always faced with the possibility of disastrous spring-time floods. The floods in these river systems fluctuated greatly in magnitude as a result of varying rates of snow melt in the mountains to the north of Mesapotamia, where the rivers have their headwaters. Because of this possibility for springtime disaster and because of the necessity to work the land during the torrid Mesopotamian summers, the civilizations which grew up between the two great rivers are characterized by pessimistic philosophies and religions based upon constant pleas for mercy from angry gods. The successive civilizations based upon Sumer and Akkad showed constant warfare among the various cities along the river bank and fluctuations in the fortunes of those cities as a result of frequent inundations during the spring floods. Additionally, gradual depletion of soil fertility through build-up of soluble salts occurred as a result of the agricultural methods utilized by the ancient inhabitants of Mesopotamia (Ref. 1-5). The social structures which developed in Mesopotamia reflected the environmental constraints of the frequent disastrous floods and the gradual depletion of soil fertility, with the resulting movement of the civilizations northward from the original foci of settlement near the southern portion of the valleys. Thus, a social change was an integral part of the adaptation to environmental change in this area. Obviously, this indicates to present day investigators that social change should be an integral part of the assessment of environmental impact in any comprehensive study efforts. With a gradual decline of agricultural potential in the Mesopotamian valleys, a concurrent gradual decay in social structure and governmental power occurred. A stronger and more virile civilization grew up to the north of Sumer, Akkad, and Babylonia, and eventually the Mesopotamian region was dominated by Assyrians and other peoples from the north.

In contrast to the situations in the Indus and Tigris-Euphrates Valleys, in the Nile Valley, the floods are very regular and do not vary in magnitude or time of occurrence in any significant fashion, from year to year. Additionally, the flooding occurs during a later portion of the year. As a consequence, the Egyptian farmer is allowed to work his fields after excess water has drained off, during the cooler winter months, and thus is enabled to harvest his crop prior to the advent of the hotter portion of the year. Although the agricultural system utilized in the Nile Valley is somewhat similar to that employed in the other great river valleys mentioned previously, the flow of water reaches a greater crest and does so in more uniform fashion from year to year, so that surplus waters may be drained off much more easily. Thus, salinization of the land, with consequent decrease in fertility, has been prevented for thousands of years. The regularity and continuity of the Nile flood is reflected in the relative stability of Egyptian civilization over a 5,000 year period in which. social customs, including rule by the Pharaohs, changed very little. Only in very recent times has man severely upset the ecological balance in the Nile Valley. This upset will be discussed in a subsequent paragraph. The contrast between the success of man in the Nile Valley for thousands of years and the eventual social and environmental failure in both the Indus and Tigris-Euphrates systems is striking.

1-3. REVOLUTION

During the so-called Dark Ages in Europe, the localized small farm was the principal mode of living, and the greatest threat to man's survival was the feudal system itself, rather than the effects of that system on the environment. However, with the Renaissance and the burgeoning industrial revolution, man's activities began to take on an increasing significance, with respect to damage to the environment. The Industrial Revolution brought an acceleration of impacts on the environment and an increase in severity of those impacts. However, the Industrial Revolution was accompanied by a scientific revolution, in which knowledge in the sciences also began to accumulate at an ever-accelerating pace.

1-3.1 Machine Slaves

The gradual deforestation of Europe has been mentioned previously (Ref. 1-1). Much of the deforestation of Europe occurred with the advent of the Industrial Revolution, as a result of accelerated demand for wood as a raw material and as a fuel. Some of the great forests, noted previously by Tacitus and other historians, disappeared prior to the onset of the Industrial Revolution, in the demand for raw materials. For example, many of the oak trees which thickly clothed the British Isles (to the amazement of the invading Roman legionary) were cut down and utilized during Elizabethan times in the 16th century for ship building. The struggle for survival of the island kingdom against Spain and the Spanish Ar-

Fig. 1-1 Newcomen atmospheric steam engine (*courtesy* Collection of Greenfield Village and the Henry Ford Museum, Dearborn, Michigan). Machine slaves: How many forests were devoured by the inefficient early steam engines which furnished power for the Industrial Revolution?

mada brought an unprecedented harvest of forest materials which even extended outside the British Isles to western Norway (Ref. 1-1). The harvesting of forestlands along the eastern shores of the Adriatic to supply wood for Genovese and Venetian galleys and merchant ships also led to deforestation, soil erosion, and depletion of the land in that area. However, with the advent of machines and their hunger for fuels, the demand for wood greatly accelerated. In later times, royal decrees were passed requiring the return of wooden barrels to the British Isles after their use in transport of exported goods, because of the relative scarcity of wood materials. With advent of the use of coal as a replacement for wood, the Industrial Revolution began a greater spurt of activity. The utilization of coal and other fossil fuels as food for the machine slaves of the Industrial Revolution led to greater and greater efforts toward extraction of other raw materials, as well as the transport and processing of those materials on an ever-larger scale. New substances were manufactured and indiscriminantly released to the environment. The extraction and transport processes had definite detrimental impacts on the environment. Additionally, the various processing operations began to release effluents into the atmosphere and into the waters of the earth in previously unseen quantities. At the time of the advent of the Industrial Revolution, the effects and significance of these effluents and impacts were scarcely appreciated.

1-3.2   Microbes, Metals, and Minds

With the Industrial Revolution came a scientific revolution and an accelerated accumulation of knowledge in the sciences, including the health sciences. Advances in microbiology and epidemiology led to the curtailment and elimination of many communicable diseases. However, efforts lagged in identifying the detrimental effects on workers of exposures to the effluents from manufacturing processes. However, these effects certainly were felt and in some trades were felt to an amazing degree. The Mad Hatter in Lewis Carroll's classic, *Alice in Wonderland,* was a caricature of an unfortunate group of individuals who developed occupational diseases through using toxic substances in the curing of beaver pelts for the manufacture of tall beaver hats. This is but one example of the extremely detrimental effects which were experienced unknowingly by industrial workers during the late 18th and early 19th centuries. With the coming of the labor movement and with further advances in the health sciences, much progress has been made in identifying occupational hazards and preventing or curing occupational diseases. However, much work remains to be done and much value can be obtained from prior investigation of proposed activities in the manufacture or processing of materials, through the completion of comprehensive environmental impact assessments for such activities. Even today, the effects on workers and on various forms of life in areas surrounding manufacturing plants, of the emission of materials such as mercury and other trace metals, is scarcely appreciated. A dramatic example of this type of toxic material which has been recently identified is vinyl chloride. It is felt that much of the injury and hazard to workers and inhabitants in regions surrounding manufacturing plants could have been eliminated through the compilation of environmental assessments for such facilities, prior to the initiation of manufacturing operations. However, even though an awakening in the health sciences was occurring concurrent with the Industrial Revolution, and even though great advances were made in the health sciences, many factors militated against the appreciation of environmental damage from man's activities during the 18th and 19th centuries.

1-3.3 The New World

One of the greatest factors which tended to blunt environmental awareness among Americans was their cultural heritage as potential conquerers of an untamed wilderness. From the very beginnings of this country, with the advent of the Pilgrims and similar groups on the eastern shores of the American continent, the national self-image for many persons was that of the heroic explorer and colonist carrying out a God-given mission to conquer a land of free air, free land, and free water. This ideology reached its most blatant expression in the early 1800's with the doctrine of Manifest Destiny, wherein the inevitable expansion of the United States to the Pacific Ocean was cited as a prime goal for all Americans. The emphasis during all of this period was on the individual and collective conquest of an untamed wilderness rather than on a conservation of precious, fragile ecosystems. It was very difficult for pioneers eking out an existence in sod huts on the Great Plains to appreciate the impact of their activities in breaking sod as possible destruction of the grasslands they earnestly desired to exploit. However, the dustbowls of the 1930's were the end result of much of this pioneer activity. The situation at the present time in the United States has drastically changed our former attitudes from conquest to conservation.

1-4. BATTLE LINES

At present in the United States, the average citizen is faced with a dilemma created by his desires for plentiful, inexpensive supplies of energy and his conflicting desire for an unsullied and beautiful natural environment. It is obvious that trade-offs must occur between the obtaining of inexpensive supplies of energy and the maintenance and enhancement of environmental quality. It is the evaluation and quantification of these trade-offs that is the very heart of environmental assessment activities.

1-4.1 The Weapons

At the present time, technology appears as the greatest threat to man's survival on earth, when viewed from the vantage point of extraction and processing of raw materials and the transportation of people and goods across the countryside. A vast panorama of destruction of natural systems can be obtained if the viewer looks through a jaundiced eye at technological activities. Technology, however, should be viewed as a powerful weapon which can be used either for good or for bad purposes. The same technology which leads to the extraction and processing of many natural materials has led to the ability to identify, evaluate, and simulate in a quantitative fashion the impacts of those same extractions and processing operations. In other words, technological advances also make it possible for man to judge the effects of his employment of technology. Technology, in this broad sense, can include means of surveillance of the earth, epitomized by the Skylab satellites and similar vehicles. Additionally, the development of digital computers has made it possible for man to accumulate, manage, and retrieve vast stores of information on the natural environment and on man's activities in relation to that environment. Finally, the application of systems science to the analysis of technological processes has made possible the quantitative evaluation of effects and impacts of technological activities on the natural environment. Thus, technology has made it possible for man to rape the land, but also to fully appreciate the fact that he is doing so and, as such, may also be used to warn man against such misuse.

1-4.2   The Battleground

One of the most controversial issues in today's world situation is the subject of limitations of the earth, which in this context, can include the limitations with respect to the amounts of fuels and other vital mineral resources necessary for industrial activity and, more important, the limitations with respect to food supplies necessary for the maintenance of life itself. There is no question that the present population of the world can be supported through proper management of the food-producing capabilities of the earth. However, the question must be asked, "At what level of subsistance will life be maintained as population increases at an ever-accelerating rate, and returns from agricultural activities begin to diminish?" In this connection, it might be pertinent to ask Americans if they are willing to accept a standard of living such as that experienced by the Chinese peasants on collective farms or the New Guinea tribesmen feeding his herd of swine in preparation for the next ritual pig slaughter. The acceptance of such a lowered standard of life is hardly tenable with most Americans today (Ref. 1-2). Therefore, the goal of impact assessment and, especially, environmental impact assessment must include an effort to maximize the standard of life for the peoples affected by the proposed activity under investigation. In maximizing this standard of living, sufficient concern must be demonstrated for coming generations, for the maintenance of long-term productivity. Therefore, the earth must be viewed as a fixed quantity with certain finite reserves of materials and mineral resources. Environmental assessments must be carried out in such a way that mineral resources may be conserved to the greatest extent possible. Additionally, environmental assessments can be the tools which will allow man to tap new sources of energy without fear of sealing his own doom in doing so. A major impetus is given to the conduct of assessments of environmental impact with every acceleration in growth of world population. This factor in itself would justify the entire environmental assessment effort now under way in the United States and in other countries. Problems still exist. Even with the application of the most up-to-date technology, misjudgments and miscalculations can be made. Incorrect value judgments may be made in assigning priorities or in judging that certain effects or benefits far outweigh the costs associated with a proposed activity. As an example of inadequate assessment of possible effects from a contemplated action, an outstanding facility is the Aswan High Dam on the Nile River in the United Arab Republic.

1-4.3 A Final Example

The Aswan High Dam was conceived as a facility to produce power and accelerate the industrialization of the developing country of the United Arab Republic. Additional benefits, through greater ability to manage the Nile River, were cited in preliminary plans for the construction of the dam. Prior to the construction of the dam, the Nile River flood occurred late each summer and produced sufficient moisture to allow farmers to plant and harvest only one crop per year, it being too dry during the remaining periods. Through the construction and operation of the Aswan High Dam, the area above the dam will have available to it perennial irrigation, and four crops may be obtained per year from the cultivated land. The benefits of industrial power supply and greater crop yield were judged, in the early 1960's, to be of far greater benefit than any costs associated with the dam itself. Even at that time, however, some individuals expressed considerable concern over the possible health effects associated with the more stable or even stagnant water conditions along the river valley (Ref. 1-10, Ref. 1-11). During the early planning operations for the Aswan High Dam, a number of health scientists and biologists indicated that there was a great possibility that the establishment of stable water conditions upstream from the Aswan High Dam would lead to a proliferation of a debilitating and sometimes fatal disease, schistosomiasis. This disease is caused by a tiny worm or blood fluke. The worms bore directly through the skins of human and animal victims and migrate to the liver. After maturing in the liver, they move to the bladder and the intestines, where they mate and deposit their eggs. The eggs are then released back into waterways through sanitary wastewater facilities. The eggs hatch in the water and become larva. The larva penetrate the body of snails found in the Nile River. In the body of these snails, the larva asexually reproduce, and multiply as much as 100,000-fold. After this reproduction in the snail, the larva emerge as free-swimming organisms to again begin the cycle of infestation in human hosts. Since the completion of the Dam, there has been a very significant increase in the infestation of inhabitants of the regions upstream and downstream from the Aswan High Dam with schistosomiasis. Because of the relatively quiet water conditions upstream of the dam, the waters also have been infested with blue-green algae. The presence of the algae and the stable water conditions produce ideal conditions for the proliferation of the snail hosts and the larva in the snails. In the stretch of river downstream from the dam for the 500 miles to Cairo, the incidence of schistosomiasis prior to completion of the dam was approximately 5%. In a recent survey in 1972, the bloodfluke incidence in river communities in the same stretch of the Nile ranged from 19% to 75% (Ref. 1-10). This is roughly a seven-fold increase in the incidence of the disease. This increased infection below the dam is the result of the proliferation of the microbial pests upstream from the dam and their subsequent migration downstream. Additionally, the utilization of perennial irrigation below the dam, as opposed to basin irrigation in former times, has created stable water conditions favoring proliferation of snail hosts and the blood flukes. In addition to the greater incidence of this serious disease among inhabitants of the Nile Valley, it has been shown that the construction of the Aswan High Dam has also led to the loss of sardine fisheries in the eastern Mediterranean Sea, increased erosion in the Nile Delta (since heavy sediment loads are no longer being carried downstream from the region above the Aswan High Dam), and a need for increased fertilization with chemical fertilizers in soils along the lower valley. No longer are these soils enriched with silt layers deposited in the September floods, as had occurred prior to the construction of the dam. This single example illustrates many of the possible dire consequences associated with construction of a major public works facility without an adequate assessment of the possible impacts thereof.

1-5. SUMMARY

In summary, the remainder of this book will be devoted to a description of the background of environmental legislation, a detailed description of the purpose and scope of the National Environmental Policy Act of 1969, an evaluation of existing methodology for impact assessment, recommendations for proper methodology, and a description of procedures for assembling an assessment team. Additionally, case histories will be presented to illustrate various facets of environmental assessment in the light of past mistakes and successes in assessment techniques.

1-5.1 The Existing Situation

At present, the need for environmental assessment becomes more serious each day. Increasing demand for energy and a conflicting demand for a clean environment require assessment of proposed activities, so that intelligent choices can be made by an informed public. Many instances could be cited to show the dire effects of rash activities undertaken without proper assessment of their environmental consequences. A growing awareness of environmental quality and the importance of that environmental quality to man's well-being further reinforces the necessity for a proper environmental assessment.

1-5.2 The Possibilities of Environmental Assessment

In addition to furnishing a means for proper evaluation of impacts from proposed actions, comprehensive environmental assessments allow a quantitative evaluation of the existing state of the environment. Establishment of existing environmental quality in some instances may be as valuable as assessing the probable impacts of a proposed action. Additionally, the compilation of a comprehensive environmental assessment may allow modifications of the design or alterations in proposed operational modes so that undesirable effects of the proposed activity may be mitigated. This opportunity to mitigate identified detrimental effects is one of the most cogent possibilities associated with environmental impact assessment. Finally, the environmental impact assessment procedure is one of the most valuable educational tools yet devised for the education of the professional and lay segments of the populace concerning environmental quality and the needs of the various sectors of the population. Too often, professionals smugly ignore public sentiments with an attitude that the supposedly uninformed public can tell them nothing. On the other hand, in many cases, members of the public act hostilely to proposed activities about which they know very little. Properly conducted impact assessment studies can do much to rectify these undesirable situations.

1-5.3 Reasons for Assessment

The overwhelming purpose for performing environmental assessment studies is the enhancement of life through the education of the public to the relative costs and benefits associated with alternative courses of action. In the end, the public must make the decision. This decision can be made through vocal protest at public meetings or through pressure mobilized through the elective process and democratic government. In any case, every individual has a right to make this choice and every individual has a right to a clean and unsullied environment. This is not a new concept in any way. In a previous paragraph, one ancient Egyptian civilization was described and analyzed. In an excavation carried out by an archaeologist probing the secrets of that civilization, a written text was found within a coffin of an individual buried approximately 4,000 years ago. This coffin text was devoted, for the most part, to a theme familiar from our own Bill of Rights, that all men are created equal in opportunity. In essence, the text stated that ". . . the sun god has made the air for all men alike to breathe, and 'I made the great inundation that the poor man might have rights therein like the great'" (Ref. 1-8). For at least 4,000 years, men have believed that poor man, rich man, beggarman, and thief have the right to equal shares in the blessings of nature. It is our belief that these men have equal rights in choosing how they will use these blessings. The goal of environmental impact assessment is to present to these men in a clear and understandable fashion the benefits and costs of all alternatives and to assist them in choosing the best alternative.

REFERENCES

1-1. Borgstrom, Georg, *Too Many,* MacMillan Co., New York, 1969, p. 4.
1-2. ibid., p. 82.
1-3. Fairservis, Walter A., Jr., *The Roots of Ancient India,* MacMillan Co., New York, 1971, p. 304.
1-4. Harris, Marvin, *Cows, Pigs, Wars and Witches: The Riddles of Culture,* Random House, New York, 1972, p. 73.
1-5. Hawkes, Jacquetta, *The First Great Civilizations,* Alfred A. Knopf, New York, 1973, p. 94.
1-6. ibid., p. 280.
1-7. ibid., p. 325.
1-8. ibid., p. 394.
1-9. Stephens, John L., *Incidents of Travels in Yucatan,* **1,** Dover Publications, New York, 1963, p. 217.
1-10. Van der Schalie, Henry, "Aswan Dam Revisted," *Environment,* **16,** No. 9, Nov., 1974, pp. 18–26.
1-11. Van der Schalie, Henry, "Egypt's New High Dam—Asset or Liability," *The Biologist,* **42,** 1956, pp. 63–70.

2

Environmental Protection, by Law

2-1. INTRODUCTION

Because of the growing capabilities to measure and monitor the effects on the quality of life from man's activities, as mentioned in Chapter 1, an increasing awareness on the part of the public, concerning environmental quality and man's degradation of that quality, has developed in recent years. This awareness is not uniform from one country to another or from one social stratum to another within a given country. However, uncertainty about man's future has grown in the minds of many people in many places throughout the world. As a consequence of this growing concern for environmental quality and a growing opposition to further degradation, organized efforts have been mounted to prevent those activities which cause environmental pollution and degradation. In both democratic and autocratic societies, laws and regulations have been passed to prevent or minimize environmental damage. This environmental protection by legislation has been only marginally successful. There are several reasons for the marginal success of these laws, and it is instructive to examine existing environmental legislation to detect its strengths and weaknesses in actually accomplishing environmental protection and conservation. For purposes of this text, only the body of environmental legislation enacted in the United States will be examined.

2-1.1 An Ethical Basis?

As mentioned in Chapter 1, a feeling of unease and uncertainty concerning the effects of industrial activity and growth has gradually arisen among the populace in many industrialized nations in the world. Although the United States had not been a leader in the development of this environmental awareness, a serious effort has been made to educate the American public to the potential risks involved in continued, uncontrolled growth, and in the continued growth of industrial utilization of raw materials and release of effluents into the air, water, and land systems surrounding polluting plants. This development in the United States has only occurred in recent decades, although the continent has been exploited for more than four centuries since its discovery.

Until the last four or five decades, the natural environment in the United States was viewed by immigrants and second-generation citizens as the subject for conquests and exploitation. Colonists and immigrants wrote glowing letters to relatives and friends left behind in their mother countries, describing the so-called "free" air, water, and land on the new continent. Although many of these letters contain references to the harsh climate and the depredations of the "savage" inhabitants, the natural fertility of the soils and the abundance of timber and other resources figured prominently in the optimistic reports concerning the early settlements, and the later expansion of the frontier, westward across the plains and into the Far West. In some ways this phenomenon was self-feeding. The optimistic reports concerning existing settlements motivated more colonists and settlers to move into the newly discovered land and to push westward in search of more fertile land. In many instances, the exploitation of the new land was ill-advised, and natural fertility was soon exhausted. A striking example of such exhaustion occurred in the tidewater areas of Virginia and surrounding states. Thus, the early colonists moved westward to escape growing population pressure, leaving depleted land to seek the paradise they had anticipated in the New World. In all of these efforts, the quality of land, air, and water was taken for granted, and no consideration was given to proper planning for the use and exploitation of these resources.

After the westward push of immigrants had been completed, with their arrival at the Pacific Coast settlements and with the completion of the frontier phase in American History, the settlers were faced with problems created by their reckless and improper use of resources. During the later 19th century, in particular, an awareness grew of the potential evils involved with unplanned development and uncontrolled exploitation. Various schools of thought were developed, and two major philosophies grew out of this effort to express national concern. One of the principal philosophic attitudes called for the preservation of large tracts of land in an untouched condition; large wilderness areas were visualized in this scheme of conservation and preservation. These untouched enclaves would be utilized by oncoming generations as living museums showing conditions as they existed when the first settlers had arrived in the area surrounding the enclave.

A second conservation philosophy developed, calling not for the conservation of enclaves, but for the controlled utilization of all land. However, in this conservation philosophy, adequate and proper comprehensive planning was considered a prerequisite to the utilization of available resources, in order to conserve and preserve those resources for the use of future generations. This philosophy, like the "enclave" philosophy, won large numbers of supporters and adherents. However, many millions of Americans were untouched by these expressions of concern and continued to support policies of uncontrolled and accelerated growth. "Bigger" was equated with "Better."

At the beginning of the twentieth century and in the years shortly following the turn of the century, greater attention and emphasis was placed upon environmental damage. During this time, much of the attention focused on damage to the individuals in the immediate vicinity of manufacturing or industrial activities, who had been harmed through exposure to the effluents of these processing facilities. With increasing identification of potential hazards from plant effluents, and with increased understanding of the interaction of all segments of the environment, the focus of attention has shifted from workers and occupational hazards to danger to the community as a whole from the exploitation and processing of raw materials.

During the last ten years, a so-called "environmental movement," generally characterized by dramatic, vocal expression of concern and fear for preservation of land, water, and air, has developed in the United States. For some time, the protection of the environment was associated with the protection of individuals' rights, and a somewhat nebulous link was made between groups dedicated to environmental conservation and groups trying to secure minority rights or to protect consumers from unfair marketing practices. However, the links among the various advocate groups have been tenuous, at best. Additionally, this intermingling of special-interest groups has somewhat clouded the issue and confused the problem of environmental conservation and preservation. At the present time, a separation of these special-interest groups is beginning to appear, with greater attention being devoted to specific areas of interest. Many environmental conservation groups now are concentrating their activities in specific areas of environmental protection and conservation and, thus, their positions are more comprehensively defined. Considerable doubt is growing in the minds of many citizens; doubt concerning the relative benefits of continued growth and continued exploitation of resources, in comparison to the consequent detrimental environmental effects associated with that growth and that exploitation. Thus, there is a significant need for assessment of relative benefits and costs, in order to allow the American public to form an educated evaluation and to formulate a national statement of environmental policy. Many people view the statutes which have been enacted in the Congress of the United States as a formal declaration of an environmental ethic, but the history of inadequate enforcement of these statutes leaves considerable doubt concerning the validity of American environmental laws as an expression of popular opinion and concern. However, statutory law can be considered a formalized expression of a national ethic, and therefore should be examined in some detail.

2-1.2 Statutory Environmental Protection

Statutory law can be distinguished from traditional law, or law arising from the customs and practices of a given people, on the basis of its formal expression in written form, and its formal enactment following the deliberations of legislative bodies. Statutory law in the United States comprises all of the public laws enacted by the United States Congress for control of both individual and governmental action. Included within the statutory laws enacted by the United States Congress for environmental protection and conservation are the National Environmental Policy Act of 1969, the Clean Air Act of 1963 and subsequent amendments thereto, the Water Pollution Control Act of 1972 and the amendments thereto, the Environmental Noise Control Act of 1972, and the Solid Waste Disposal Act of 1965, as amended by the Resource Recovery Act of 1970, and the Resource Conservation and Recovery Act of 1976.

The references to this chapter contain a comprehensive list of current and prior laws enacted to obtain environmental protection. These laws constitute the formal expression by the federal government of the national concern for environmental quality, and are attempts to protect and preserve the environment through legislative action. Obviously, legislative action will be ineffective if the laws which have been passed are not enforced. Therefore, some attention should be given to the administrative procedures and policies which have been promulgated within the federal government to enforce environmental legislation. Later in this chapter, a review of agencies and their responsibilities within the Executive Branch will be given.

2-1.3 Common Law

In addition to statutory regulation of environmental pollution, the traditional laws based upon Anglo-Saxon customs and practices also can be utilized to obtain environmental protection. Common law, as opposed to statutory law, has not been written in precise form except in regard to those Acts of Parliament which were in force in the original thirteen colonies at the time of the American Revolution. The basis for common law, through the years, has been the interpretation of customs and practice by the courts of the land. In other words, violation of common law has been seen by the Judicial Branch of the government, and, in particular, by local courts, whenever a particular action has been shown to be contrary to the customary practices of the American populace, considered through the historical perspective of the common law which existed at the time of the revolution, as well as subsequent court rulings and decisions concerning common law. The weight of tradition and previous court decisions is brought to bear in common law cases against an alleged violator of traditional practice. With respect to environmental protection, traditional guarantees of individual rights and traditional definitions of wrong-doing have been utilized to stop or prevent environmental degradation.

The three principal areas in which the common law has been invoked in environmental cases have been the areas of negligence, nuisance, and trespass. These three categories of offense have been utilized in subsequent litigation. However, in many situations, especially with regard to noise pollution, the doctrine of nuisance has been utilized to stop the generation of excessive noise at particular times. The utilization of common law, in general, has occurred whenever the statutory provisions for environmental protection have failed to achieve the desired results.

2-1.4 "Case" Law

With respect to both statutory provisions and common law provisions, the enforcement of environmental protection and preservation has depended upon the interpretation of both types of laws in the courts. In other words, the judicial interpretation of statutory law can either handcuff or fully reinforce the actions of the law-makers and those who would enforce the laws. In essence, since the duty of the judicial branch of government is to interpret and expound the law of the land, the law is what the courts say it is (Ref. 2-1). Case law is very instructive, since a review of court decisions can indicate other facts in addition to the assessment of liability in a particular case. At times, court decisions decide basic issues involved in certain statutory laws. For example, in the case of *Calvert Cliffs Coordinating Committee vs. Atomic Energy Commission* (449 F.2d 1109), a basic issue, with respect to the National Environmental Policy Act of 1969, was decided. In that case, the court ruled that each environmental assessment must be carried out on an individual basis, from one project to another, so that the balancing of benefits and costs, in terms of the natural environment, could be done so that for each individual case, an optimal course of action could be identified. In other decisions, the court has given binding interpretations of somewhat ambiguous laws or statutes. For example, in the case of *Environmental Defense Fund vs. Corps of Engineers* (325 F. Supp. 728), the ambiguous character of the National Environmental Policy Act of 1969 was clarified, setting specific requirements concerning the contents of Environmental Impact Statements. In this case, the court ruled that the ambiguous language of the National Environmental Policy Act should be construed to indicate that a systematic, interdisciplinary approach for environmental assessment is required in all cases. In other instances, court decisions have decided legal points with respect to individual rights to due process. In the case of *Sierra Club vs. Morton* (405 US 345, 2 ELR 20192, 3 ERC 2039), the Supreme Court gave a major decision concerning "standing," or who may bring suit to obtain environmental protection. Additionally, judicial review of agency actions can be sought, and such review can constitute binding restrictions on further agency actions. For example, in the case of *Scherr vs. Volpe* (336 F. Supp. 882), the court found that a federal agency had actually violated its own guidelines in several of its actions. The power of the court in pointing out this discrepancy in agency action was significant in governing the future agency activities. Finally, studies of case law are important since court decisions indicate trends in judicial interpretation and, therefore, can form a basis for the prediction of future decisions. Of significant use in attempting such predictions are the comments frequently appended to the terse statement of decision in a court opinion. The lengthy statements made by judges in rendering the court's decisions in particular cases are not without value in that they indicate the probable direction of future decisions by members of that particular court or members of lower courts. (Ref. 2-7 contains a very comprehensive summary of environmental case law for the United States.)

2-2. COMMON LAW

As mentioned previously, the term "common law" designates that body of legal theory and law which was formulated and developed in England and was in force in the United States at the time of the American Revolution. This law is based solely on the customs and usual practices of the community and does not derive its authority from a written statement based upon legislative action. In essence, common law is the ancient unwritten law of Anglo-Saxon England. It has been developed over the years through successive restatements in the form of court opinions reaffirming certain points in the law and restating opinions given in earlier cases. In common law cases, much emphasis is placed upon precedents. In the United States, the common law is based upon that of England, with only one exception; in Louisiana, the common law is based upon ancient Roman law, since the state of Louisiana was under French rule at the time of its purchase during the Administration of President Jefferson, and the French Napoleonic Code was based upon Roman law. The importance of common law in environmental protection is not insignificant. Particularly with respect to pollution control, utilization of common law has produced worthwhile results. As stated above, the three common law actions which have most frequently formed the basis of lawsuits have been negligence, nuisance, and trespass.

2-2.1 Negligence

The term "negligence" pertains to a situation in which one party has caused injury to a second party or to a second party's property by acting contrary to, or failing to act in accordance with, some legal duty owed by the first party to the second party. By definition, the action or inaction of the first party is intentional, but the resultant injury to the second party or his property is unintentional. One of the most important considerations in deciding questions of negligence is the degree of care or the standard of performance (i.e., the legal duty) which should be required by law in a particular situation. In general, the degree of care required by law is that degree of caution and prudence which would be exercised by an ordinary individual of average intellect and prudence (a "reasonable man") under similar circumstances. If it can be demonstrated that the defendant had a legal duty toward the plaintiff and that a reasonable man, under circumstances similar to that of a given case, could have forseen the undesirable, duty-breaching effects and the injury to the plaintiff, then the defendant generally is found to be guilty of negligence. However, in order for the defendant to be liable for the damage created through his negligence, it must be shown that the negligence was the proximate cause of the detrimental consequences or injury. Generally, if it can be shown that in the normal course of events, the injury would not have occurred if the negligent omission or negligent act had not taken place, then the defendant is held to be liable for the consequences of his negligence.

In some instances, it is very easy to demonstrate negligence. For example, if a statute or law passed to promote safety and minimize hazards was violated, then the violation constitutes negligence. This is "negligence *per se*." However, the negligent defendant will not be liable for the damages which occurred unless it is shown that the negligence (the violation of the safety law) was the actual and proximate cause of the damage or injury. Additionally, the violation of a safety standard does not constitute negligence for which a defendant is liable if the plaintiff or the injury resulting from the violation are not within the "zone of interests" for the protection of which the statute was enacted. In other words, if a defendant had violated a safety code by not providing water sprinklers in an industrial plant, as protection against fire, this violation of a safety requirement would not make that defendant liable for injuries suffered by a plaintiff who was injured when a boiler exploded as a result of safety valve malfunction. Additionally, a defendant may not be liable for damages incurred in a situation in which a plaintiff's negligence contributed to the injury or to the hazardous situation. Under the principle of "contributory negligence," if the plaintiff had the opportunity to avoid the injury and did not do so as a result of his own negligence, the defendant will not be found liable for damages, unless it can be shown that the defendant had, and did not utilize, the last clear chance to prevent the injury and its consequences.

In some instances, defendants are held to be strictly liable for the results of their actions. For example, in the environmental protection field, release of hazardous substances from a land disposal site can be considered to be behavior for which the defendant operator of the land disposal site can be held strictly liable. In this case, the deposit and storage of potentially hazardous materials at a particular site is considered to be a guaranteed, or certain, cause of injury to those exposed to the hazardous substances should they escape. This doctrine of strict liability has been applied in several cases where chemical effluents or petroleum products have escaped from containment facilities, entering streams and watercourses. In the environmental protection field, it has, in the past, been very difficult to prove the defendant's negligence, because of difficulty in establishing a cause-and-effect relationship between the alleged negligent actions and the consequent environmental damage or injury to the plaintiff. However, several recent court decisions indicate that the failure to utilize the *best current practice* in pollution control technology will, in the future, be considered to constitute negligent behavior. The provisions of the Clean Air Act Amendments of 1970 and the Federal Water Pollution Control Act Amendments of 1972 call for increased inspection and monitoring of point sources of air and water pollutants. The gathering of comprehensive source emission data, required under the provisions of these and other new statutes, may create a body of data from which conclusive evidence of negligence can be deduced. Thus, in the future, the utilization of the negligence provisions in common law should increase in attempts by the public to achieve environmental protection and redress from environmental degradation.

2-2.2 Trespass

Trespass can be defined very generally as any physical invasion, by one person or his property, of another person or his property, which causes an injury to the person, property, or rights of the second individual. Trespass to the person of another individual is some unlawful action committed directly on the person of the individual involved. Trespass to personal property is any action which interferes with the possession of personal property on the part of another individual, with or without direct physical action. Trespass to real estate is an unlawful entry onto another's property. The act of entry onto real estate constitutes the trespass, and the trespass does not depend upon detrimental effects from such entry. Thus, trespass can occur regardless of the condition of the land entered and regardless of the negligence of the landowner in preventing entrance. In all of these instances of trespass, a defendant is liable for any damages created through the trespass even if the defendant acted with reasonable prudence and best current practice. In the environmental protection field, trespass actions have been brought most frequently for trespass to real estate. In these actions, the plaintiffs have claimed that the defendants have caused or permitted a thing or things to intrude upon the real estate of the plaintiffs. This application of common law is particularly effective with respect to the measurable and visible emissions of air pollutants and water pollutants from point sources and their transmittal to adjacent land owned by another individual. Trespass on real estate has been applied successfully to control emissions of air and water pollutants, but has not been used successfully in the abatement of noise pollution. This distinction is based upon the criterion that trespass occurs only through an actual physical invasion of real estate by *material* items. Wastewaters or constituents in wastewaters, and air pollutants, certainly can constitute material transmitted onto real estate in trespass cases. The action of intrusion of these items onto the real estate is sufficient to establish a trespass action; in trespass cases, no actual damage or detrimental result need be shown from the intrusion of material onto property. The provisions of the 1970 Amendments to the Clean Air Act include a consideration of trespass, in that invasion of property by air pollutants is clearly defined as constituting trespass. Consequently, trespass actions brought under provisions of the 1970 Amendments to the Clean Air Act should increase in frequency and importance in coming years.

2-2.3 Nuisance

In addition to claims of negligence under provisions of the common law, environmental protection has been sought through the law of nuisance. Nuisance is defined in Black's Law Dictionary (West Publishing Company, St. Paul, Minn., 3rd edition, 1933, pg. 1738) as that ". . . class of wrongs that arise from the unreasonable, un-warrantable or unlawful use by a person of his own property either real or personal, or from his own lawful personal conduct working an obstruction of or injury to the right of another or of the public in producing material annoyance, inconvenience, discomfort, or hurt." Nuisance is distinguished from another aspect of common law, trespass, in that nuisance involves the use of one's own prop-erty in such a manner as to cause detrimental effects, whereas tres-pass constitutes a violation of property rights of a second party.

   In general, there are two types of nuisances, private and public. A public nuisance is one which affects the community at large, or through which an indefinite number of persons are injured. On the other hand, a private nuisance is a condition in which a limited number of individuals (or a single individual) are injured. In gen-eral, in order for injured parties to obtain relief from a public nui-sance, a lawsuit to stop or prevent the nuisance must be brought by a public official or by a private citizen who can show some injury greater than that suffered by the general public. In only a few cases the courts have decided that individuals could recover damages when public nuisances were shown to be the causes of the damage. In private nuisance cases, the most common form of action is the granting of an injunction, on behalf of the injured individual, to halt the detrimental action and remove the nuisance. In general, the ef-fluents from a point source, whether they be air pollutants, water pollutants, or noise, are considered to be a nuisance if they in-juriously affect the health and/or comfort of ordinary individuals living in the vicinity of the point source; i.e., the polluter's action unreasonably interferes with the plaintiff's use and enjoyment of his property. Because of the judgments involved in deciding who are ordinary people and what is an unreasonable extent, there is no fixed standard in nuisance cases and each decision rests on individ-ual consideration of the merits of a case by a judge or jury.

In the environmental protection field, the most common nuisance actions are asociated with odor or noise. Here the nuisance action has been a great aid to environmentalists, whose early efforts under trespass law were stymied by the impossibility of showing a physical invasion by "noise" or "smell." In the past, it had also been difficult to prove the emissions of physical air pollutants constituted nuisances, because of the difficulty in differentiating between polluted air and so-called normal air in a given locality. This situation has been rectified, to a great extent, through the establishment of primary air quality standards in the Clean Air Act of 1970. Air pollutant concentrations which exceed the standards in the Clean Air Act are considered to constitute definite hazards to the health and well-being of persons exposed to these pollutant levels. Thus, emissions of air pollutants in concentrations or degrees greater than those stipulated in the primary standards for air quality in the Clean Air Act of 1970 clearly can be construed to constitute nuisance situations. With regard to noise pollution, the definition of noise levels at which a nuisance is created has been very difficult in the past. It has been especially difficult to prove that the generation of noise intensities of a given magnitude has seriously and unreasonably injured the health and well-being of ordinary persons living near the noise source. However, passage of the Occupational Safety and Health Act and the Noise Control Act of 1972 have clarified this situation to a great extent. In the Occupational Safety and Health Act, decibel limits were established, and generation of sound pressure levels in excess of the limits given in this Act, for periods of time in excess of those also given in the Act, clearly constitutes a hazard to the hearing and psychological well-being of persons subjected to those noise levels. Thus, the Occupational Safety and Health Act and the Noise Control Act have established scientific guidelines and criteria for the evaluation of exposure to noise of a given intensity for a given amount of time. Through the use of these guidelines, it is possible to determine what is a physically and psychologically dangerous noise level under various circumstances. This determination of clear hazard corresponds to the determination of "injury to an unreasonable extent" embodied in the definition of nuisance. It appears likely that the establishment of standards for water quality in various regions, as part of the 1972 Amendments to the Federal Water Pollution Control Act, may also lead to a clear definition of nuisance in the water pollution field, and that nuisance actions may grow in number and importance in water pollution control efforts in coming years.

2-2.4 Defenses Under Common Law

A number of different positions have been taken by defendants in civil lawsuits brought under provisions of the common law. These defenses range from claims of contributory negligence, to claims of public authorization for the supposedly detrimental activities. As mentioned previously, one of the easiest defenses for a defendant in a nuisance action is to confuse the issue by maintaining that if a nuisance has been created, it is a public nuisance and, therefore, action can ordinarily be taken only by a public official on behalf of the large number of persons affected. In other instances, defendants in nuisance actions have claimed that the plaintiffs have been aware of the supposed nuisance activity, and that the knowledge of the alleged nuisance on the part of the plaintiff constitutes a defense against, or waiver of, any such claims. For example, in many noise pollution cases where airport noise is alleged to be a nuisance, the airport officials have attempted to show that the persons bringing the nuisance action were aware of the noise created by the airport prior to the time they purchased property near the airport. A few courts have held that such previous knowledge of a given situation constitutes a defense against claims of nuisance, but the majority of courts are now rejecting this theory as a defense in a "pollution" lawsuit.

The major defense in negligence cases, as mentioned previously, has consisted of the claim of contributory negligence on the part of the plaintiff. In other words, negligent conduct on the part of the plaintiff has contributed to the hazardous situation and has resulted in injury to the plaintiff for which no liability should be imposed upon the defendant. In addition, if it can be proved that a plaintiff knows of a risk in advance and assumes that risk, any claim for damages on the part of that plaintiff generally can be defeated. In many of the negligence actions which have been brought in the environmental protection field, the defense takes the form of a denial of any causal relationship between a supposedly negligent action and the detrimental consequences suffered by the plaintiff or plaintiffs. As stated earlier, the use of this defense has been greatly reduced, through requirements of comprehensive monitoring activities in the amendments to the Clean Air Act and the Federal Water Pollution Control Act.

Other defenses have been put forward in civil lawsuits asking redress for environmental degradation. One of the technical legal defenses most frequently used has been laches. This term denotes a failure, on the part of a plaintiff, to enforce or claim his rights at a proper time, or to perform a required action at a proper time. In essence, the defense in such situations arises from alleged acquiescence, on the part of the plaintiff, in the acts which caused the environmental damage and pollution. This defense has been used successfully to prevent the injunction and halting of work already far in progress before belated lawsuits are brought by plaintiffs opposed to a particular action.

In other cases, defendants have claimed that the action which has caused environmental pollution and degradation has been authorized either through contract with a second party or through public authorization for the alleged polluting activity. In general, if a particular plant or source of effluents has been licensed or given administrative approval by regulatory agencies, such legislative or administrative approval has been considered to be an adequate defense against claims of liability for resultant pollution. However, any legislative license to allow pollution and, therefore, to allow the creation of a nuisance, must be explicitly expressed. Even where a point source or plant has been given explicit legislative sanction, claims of private nuisance may arise and may be considered justifiable. In such situations, defendants may be considered liable for damages to private individuals if the existence of a private nuisance can be demonstrated.

Finally, defendants in pollution cases have sought to prove that regulatory statutes or codes are too vague for proper enforcement, or that the statutes and regulations governing pollution control are unreasonable and impossible. This has been one of the major claims of the automobile industry, with respect to the installation of air pollution control equipment on new automobiles. The major auto makers have claimed that it is impossible to manufacture automobiles with pollution control equipment sufficiently effective to meet air quality standards stated in the Clean Air Act Amendments of 1970. However, such claims have met with very little success in the courts. A few isolated instances have arisen where courts have ruled that regulations requiring pollution control are unreasonable because the technology adequate to meet the control standards does not exist, and that the defendant party could not possibly exercise the required control, even through the best currently available technology. As environmental quality standards are more explicitly stated in new laws, claims of vagueness will diminish. Additionally, with continuing development of control technology, defenses based upon the alleged unreasonableness of regulatory standards should also disappear. However, this is one of the most controversial sectors of environmental law and remains undecided at the present time. Because of the changing scope and capabilities of pollution control technology, it is virtually certain that continued interpretation of the phrase "best current practice" in pollution control efforts will be required on the part of the courts.

2-2.5 Summary

In summary, a significant amount of effort has been expended, attempting to obtain environmental protection through provisions of the common law handed down through the years from Anglo-Saxon England to the American Colonists and to later generations of Americans. The three most common actions brought in pollution control attempts under provisions of the common law are suits in negligence, nuisance, and trespass. In general, it has been difficult to prove negligence in pollution control cases because of the difficulty in showing a causal relationship between the release of pollutants and the detrimental effects of such release. Increasing capabilities to monitor pollutants and to monitor community health around point sources of pollutants, and increasing emphasis on the maintenance of emission records, should do much to clarify the situation with respect to negligence. Claims of nuisance have been made primarily in connection with noise pollution and air pollution cases. It appears that with respect to noise and odor pollution, the claims of nuisance offer the best possibility of redress for environmental damage. Definitions and evaluations of noise levels harmful to human beings, contained in the Occupational Safety and Health Act, can serve as the basis of proof in many nuisance cases. Trespass claims have often been utilized in the past to stop or prevent the emission of pollutant materials from point sources onto adjacent lands. It is likely that in the future, trespass cases will continue to form a significant part of the legal effort to prevent environmental pollution. The common defenses against charges of negligence, nuisance, and trespass have rested upon claims that the injured individual has contributed to the detrimental situation and therefore has been partly responsible for the injury, or that the person has been aware of a nuisance and has placed himself in jeopardy while aware of the potential hazard. Other defenses which have been put forward are claims that new regulations and statutes are impossible to enforce, are unreasonable, or are too vague to interpret correctly. With the advent of more explicit and detailed environmental statutes in recent years, these claims of vagueness and unreasonableness have decreased significantly in number and in importance. A continuing controversy centers around the possibility of compliance with new environmental control legislation. It is likely that such controversy will continue in the future and will intensify as a result of the growing economic impact of pollution control activities.

2-3. STATUTES FOR ENVIRONMENTAL PROTECTION

In addition to the body of common law which has been handed down over a long period of time and reinforced through new judicial interpretations, a significant body of legislation has been enacted to give formal expression to a concern for environmental quality and a concern for environmental protection in the United States. The federal laws directed toward obtaining environmental protection and preserving environmental quality are discussed briefly in this section.

2-3.1 The National Environmental
Policy Act of 1969

The National Environmental Policy Act of 1969 (NEPA) is the most important piece of federal legislation designed to achieve environmental protection and preservation. Because of the influence of this Act on all federally-supported activities, and because of the requirements within the Act for comprehensive assessment of environmental consequences of proposed activities involving federal funds, it is described in the next chapter in a separate presentation. In brief, this Act was passed to insure that all federally-supported activities will be undertaken only after the completion of comprehensive planning efforts which include sufficient safeguards for existing environmental quality and the preservation and enhancement of future environmental conditions. The history, development and scope of this Act are presented in Chapters 3 and 4, with the methodology for achieving assessment as required by NEPA given in Chapter 5. Ref. 2-27 contains the full text of the National Environmental Policy Act of 1969; the text of this law is also given in the appendix of this book.

2-3.2 Air Quality Protection

Air pollution control legislation was introduced into the Congress more than 25 years before the writing of this volume (Ref. 2-2). However, no law was passed by Congress until 1955, when Public Law 84-159 was enacted. This law authorized the expenditure of $5,000,000 per year for a 5-year period through 1960, for study and research on ambient air quality in the United States, and on the means for air quality control. During this time, the Executive Branch was opposed to federal intervention in business activities and, in general, the enacted legislation thus was ineffective in producing any significant abatement of air pollution or in creating any improvement in national air quality.

During the early years of the 1960's, with a change in Administration, a new effort was put forward for the prevention of air pollution through effective legislation. In 1963, a Clean Air Act was passed and signed into law by President Kennedy: Public Law

88-206. The chief thrust of this law was to enable the Department of Health, Education, and Welfare to conduct public hearings concerning alleged incidences of air pollution. However, the hearings could be conducted only at the request of individual state governments. If the hearings did not result in an abatement of the air pollution under investigation, the Department of Health, Education, and Welfare was further empowered to call a general conference on the air pollution problem at hand. If this general conference was not successful in producing an abatement of the identified air pollution, the DHEW could request a federal court action in the form of an injunction against the alleged air polluter. Because of the sequential nature of actions called for in the 1963 Clean Air Act, and because of the long time delays associated with each step in this procedure, the effectiveness of the Act was practically negligible. The cumbersome procedure included in the Act made it virtually unenforceable.

One of the major sources of air pollutants in the United States, the internal combustion engine, was identified early in the program of study of air pollution in this country. Action was not taken against this particular source of pollution until the passage of the Motor Vehicle Air Pollution Control Act, in 1965. This Act empowered the DHEW to establish emission levels for new motor vehicles. The Department was enabled to establish permissible levels of pollutant emissions, however, only for engines and ventilation systems for new motor vehicles manufactured after 1965.

The next major step in the legislative battle against air pollution was taken with the passage, in 1967, of major amendments to the Clean Air Act of 1963. These amendments authorized the DHEW to develop air quality standards for the entire United States, and to enforce control measures designed to maintain proper levels of air quality as identified in the first part of the standard-setting procedure. The 1967 amendments also were relatively ineffective since they included a provision for an intergovernmental system for air pollution control on a regional basis. The procedures included in the amendments called for the DHEW to designate air quality control regions, to establish ambient air quality criteria within those regions, and to develop reports on control techniques and devices for air pollution control. The amendments simply urged the individual states to adopt air quality standards in line with those developed by DHEW and to subsequently implement plans to limit point-source emissions in accordance with the adopted standards. Because of the weak provisions in these amendments, the overall effect of the 1967 act was to delay any effective action toward air pollution control and improvement in air quality. Between 1963 and 1970, for example, under the provisions of the original Clean Air Act, only ten enforcement conferences were held. Four of the conferences were concerned with emissions from individual point sources and the remaining were concerned with air quality in large regions. As a result of this activity, in only one case of air quality violation was a request for federal court action made. In this case, *U.S. vs. Bishop Processing Co.*, even after 14 years of procedural delays, and compliance with the cumbersome mechanics of the 1963 Act and the 1967 Amendments, less than a complete prevention of further emissions was achieved.

Most of the failings of the previous legislation were remedied in 1970, when the Congress passed further Amendments to the Clean Air Act as Public Law 91-604. These amendments contained the first provision for federal intervention to protect and preserve a segment of the environment. Corresponding legislation to prevent water pollution was not enacted until 1972. In the 1970 Amendments to the Clean Air Act, Title I authorized the Administrator of the Environmental Protection Agency to designate air quality regions based not only on political boundaries and zones of urban development, but also upon considerations of climate and topography. This basis was far superior to the guidelines set forth under prior air quality statutes. At the present time, more than 250 air quality regions have been established in the United States. The 1970 Amendments also required the Environmental Protection Agency to develop criteria for air pollutant concentration levels, in order to determine the level of pollutants at which adverse affects could be produced in human beings and in other living organisms. On the basis of the concentration levels so identified, the EPA was further authorized to establish ambient air quality standards based upon "primary" and "secondary" effects on living segments of the environment. In this context, primary standards are those which are necessary to prevent direct effects of air pollution on members of the public, and secondary standards are those designed to protect the public from any indirect effects of air pollutant release. The current standards for primary and secondary air quality are contained in the air quality laws given in Refs. 2-12 to 2-19. In addition to the foregoing authorizations, Title I of the 1970 Amendments also called for the development of implementation plans for air quality preservation on the part of individual states. The Environmental Protection Agency was directed to supersede state activities if the state implementation plans were judged to be inadequate in light of national or regional maintenance of air quality and control of air pollutant emission. Additionally, Title I of the Amendments directed that within any given air quality region, no deterioration of existing air quality should be allowed. All new stationary sources of air pollutants are required to be equipped with the best obtainable air pollution control devices at the time of their construction. In order to reinforce state air pollution control activities, the federal government was empowered to enforce the emission control regulations developed in each state. In the event of a violation, the EPA is required to notify the state in whose jurisdiction the violation occurs, and to notify the violator. If the air pollution is not abated as a result of the notification, the Environmental Protection Agency can institute civil action against the violator or can implement stricter regulations to supersede the state authority. Under provisions of the 1970 Amendments, fines of up to $25,000 per day and/or one year in prison can be leveled for continued emissions of air pollutants, for an initial conviction. A second conviction for continued air pollution can result in a doubling of these penalties. On the other hand, with the exception of motor vehicle emissions, state governments were empowered under provisions of the 1970 Amendments to implement even stricter control than the federal regulations, if they so desired.

Other activities called for in the 1970 Amendments included the requirement for operators of stationary sources of air pollution to monitor the generation and dispersal of such air pollutants, to maintain emission records, and to make those records available to the public upon request. As mentioned in a previous section, this provision may be quite effective in any attempts to prove negligence on the part of an air polluter. Finally, under provisions of

Title I of the 1970 Amendments, the Environmental Protection Agency was empowered to establish emission standards for hazardous substances. The full impact of this clause in Title I has yet to be felt, but it should be extremely significant in the limitations of pollutants from chemical processing industries.

Title II of the 1970 amendments was devoted to the control of emissions from mobile sources, particularly motor vehicles. The provisions for control of auto emissions, as established under the 1965 act, were greatly strengthened under Title II of the 1970 Amendments. The authority to establish emission standards was transferred from the Department of Health, Education, and Welfare to the EPA. Title II of the Amendments further required a 90% reduction in carbon monoxide and hydrocarbon emissions by 1975, and a 90% reduction in nitrogen oxide emissions by 1976. In 1971, the EPA established the baseline data for determination of the appropriate reductions and stated that carbon monoxide must be reduced from approximately 34 grams per vehicle mile to 3.4 grams per vehicle mile. Hydrocarbon emissions were to be reduced from 4.1 to 0.41 grams per vehicle mile, and nitrogen oxide emissions were to be reduced from 4.0 to 0.4 grams per vehicle mile. Stricter testing of new vehicles also is required under Title II, and the Environmental Protection Agency is enjoined from issuing verifications of compliance unless the vehicles comply with the new standards in all aspects. The manufacture of a new motor vehicle without verification of compliance is to be considered a criminal action and penalties of up to $10,000 per violation were prescribed. The 1970 Amendments further empowered the EPA to obtain restraining orders or injunctions from federal courts to prevent continued air pollution when a clear and imminent danger to public health could be demonstrated as a result of the alleged air pollution. Furthermore, the Federal Government is prevented from purchasing products or services from individuals or facilities which have been convicted of air pollution violations under Title I of the 1970 Amendments. Finally, because of special local conditions in that state, California was exempted from the provision for federal intervention in regulatory activities concerning motor vehicles, in order to allow that state to establish and enforce standards even stricter than the federal standards.

In general, the air quality control legislation which was passed prior to 1970 was largely ineffective because it attempted to achieve air pollution control through the control of atmospheric constituents (Refs. 2-15 to 2-19). This is neither legally nor technically possible. On the other hand, the 1970 Amendments are aimed at controlling emissions of specific pollutants from particular sources. These amendments can be effective in limiting emissions even in situations where the air quality of an entire region has not been defined. The provisions for monitoring of state implementations plans by the Environmental Protection Agency also is a significant improvement over prior legislation. This provision allows the EPA to require effective controls in areas where local vested interests may militate strongly against the establishment of air pollution control. The development of comprehensive air pollution control through legislation and regulation will be significant only if the EPA is effective in managing the activities with which it is charged under the authority of the 1970 Amendments. Although some progress has been made to date, some concern has been expressed recently because of a seeming slowdown in enforcement activities on the part of the EPA. Public sentiment, in part, favors further emission of air pollutants from motor vehicles in response to decreasing availability of low-sulfur fuel. A clear conflict has developed between the maintenance and enhancement of air quality under provisions of the 1970 Amendments and the maintenance of a low-cost, plentiful supply of energy. The so-called energy crisis has had a significant effect upon the public concerned with maintenance of air quality. In the face of possible allotment and rationing of petroleum resources, many individuals feel that other fuels, such as coal, should be burned to produce energy, regardless of the environmental effects thereof (Ref. 2-33).

On June 22, 1974, Public Law 93-319 became effective as the Energy Supply and Environmental Coordination Act. This Act was aimed at encouraging energy conservation and conversion to use of coal rather than oil or natural gas. This law is very ambiguous and inconsistent with existing laws but still may have serious effects on air pollution control efforts. It is, in part, a set of amendments to the Clean Air Act and partially a new law. In the amendments section, the Administrator of the EPA is given the authority to suspend air pollutant emission limits for plants forced to utilize coal in lieu of other fuels. Section 3 of the new law allows the EPA to over-rule state emission reduction programs and allow point source dispersion of pollutants via high stacks. This provision is in conflict with existing legislation and also may create hostility between State governments and the EPA. Furthermore, this section allows deterioration of ambient air quality—a serious conflict with the 1970 Amendments and several Supreme Court decisions. Other potential conflicts are created in other sections of the Act which (1) allow the Federal Energy Administrator to prohibit the use of fuels other than coal in given plants, (2) give the EPA power to veto actions of the FEA, and (3) give the FEA power to allocate coal, which could effect EPA activities under the Clean Air Act.

Also, comprehensive reports to Congress are required from the EPA and the FEA on energy supplies, mass transportation, transportation air pollutant emission control, and effects of the energy/environment conflict.

This ambiguous and contradictory law is sure to cause the filing of a large number of lawsuits by power producers, environmental advocate groups, and others. It is likely to do little good in solving the problem of low-cost energy in a clean environment.

It is pertinent to point out at this juncture that local air pollution control achieved through the passage of local statutes or ordinances should not be disregarded, even though the major authority for air pollution control in the United States has been given to the Federal Government. Under provision of the 1970 Amendments, the Federal Government was empowered to supervise enforcement of air pollution control regulations, but provisions also were made available for local participation in improving air quality within given regions. For the most part, the most effective local control of air pollution has occurred in those highly industrialized and densely populated urban areas of the United States where air pollution effects have been most severely felt. The most successful efforts for air pollution abatement have been based upon ordinances designed to achieve emission control from point sources. The issuance of permits for operation of facilities only after installation of air pollution control equipment, has proven to be an effective technique for obtaining air pollution control in many cities. The success of these efforts should not be disregarded.

2-3.3   Water Quality Protection

Federal laws pertaining to the maintenance of water quality actually have been in existence for more than 75 years, but only within the last ten years has meaningful legislation been enacted. In 1912, the United States Public Health Service was authorized by Congress to investigate pollution of the national water system. This authorization was little more than "lip service" to the concept of water pollution control and no significant work was carried out during the first 40 years of this century. In 1948, as a result of mounting pressure from concerned citizens, Congress passed a weak water pollution control act which was virtually unenforceable. This act required the Public Health Service to assist individual state governments to obtain water pollution control through the provision of technical advice and information on control techniques and equipment. A Water Pollution Control Division was established within the Public Health Service, and a very small amount of money was allocated for the construction of sewers.

In 1956, Public Law 84-660 was enacted to allocate some funds for research on water pollution. Funds also were granted to match any available funds held by municipalities for the construction of municipal wastewater treatment plants. However, the administration of the Executive Branch at this time, as mentioned previously, was generally opposed to federal intervention or spending to achieve pollution abatement; such intervention was considered to be the province of the individual state governments involved. Thus, little effective work was done under provisions of the 1956 law.

In the 1960's, increasing pressure was brought on Congress by a new Administration to change and strengthen federal water pollution control legislation. With the passage of the Water Quality Act of 1965, Public Law 89-234, the Federal Water Pollution Control Administration was created within DHEW, and serious federal activity to prevent water pollution was finally initiated. The 1965 Water Quality Act increased federal funding of construction grants for wastewater treatment facilities, and also required state governments to prepare stream quality standards and control standards for pollutant emissions into state waterways. In 1966, the Federal Water Pollution Control Administration was transferred to the Department of the Interior.

In 1970, the Water Quality Improvement Act was passed as Public Law 91-224. This law strengthened the control on emissions of waste materials from ships and other navigable craft, and also provided stricter control measures for oil spills. The Federal Water Pollution Control Administration was renamed the Federal Water Quality Administration, and a short time later, in 1970, was transferred into the newly-created Environmental Protection Agency. After the creation of the EPA, a rather unique situation developed in response to demand for a meaningful enforcement of federal programs for water pollution control. The 1899 Rivers and Harbors Act, which had delegated control over pollutant and waste dump-

Fig. 2-1 Dam and locks under construction on a major navigable river (*courtesy* Louisville District Corps of Engineers, Department of the Army). Early environmental legislation delegated the protection of waterways and navigation to agencies such as the Corps of Engineers, Department of the Army.

ing into navigable waterways to the US Army Corps of Engineers, was resurrected as a means to achieve water pollution abatement. This act was originally designed only to ensure the maintenance of navigation capabilities on waterways, but a broad interpretation of the original act by a federal court gave the Corps the authority to prevent water pollution in navigable waterways by exercising control over any material introduced to such waterways. This tenuous extension of authority of an antique law did not prove effective in obtaining comprehensive water pollution control, and further action was taken by Congress in 1972, with the passage of Public Law 92-500.

A review of existing laws passed prior to 1972, and a comparison with the Federal Water Pollution Control Act Amendments of 1972, will show that these amendments completely revamped the federal approach to water pollution control (Ref. 2-3, 2-20, 2-21, 2-22, 2-23, 2-24, 2-25, 2-26). The Amendments established two goals in water pollution control activities: the establishment, by 1983, of waterways sufficiently pure for swimming and other recreational uses and for the propagation of fish and wildlife; and, by 1985, the complete elimination of all pollutant discharges into the nation's waters. The 1972 Amendments were radically different from prior laws in two significant ways. First, the Federal Government was empowered to accept or reject the Water Pollution Control plans drawn up by individual states. Additionally, the Administrator of the EPA was authorized to supersede state rules and regulations and to directly control water pollution control activities within the boundaries of any state which did not comply with the national program for water pollution abatement. Secondly, the Federal Government was authorized under the 1972 Amendments to obtain immediate court action against any polluters of national waterways if the actions of such polluters could be shown to present a clear and imminent danger to public health.

In addition to the provisions mentioned above, the 1972 Amendments also provided funds for research, including investigations of ambient water quality conditions, the establishment of a water quality surveillance system, and the development of new water pollution control techniques directed toward non-point sources of water pollution. Over $200,000,000 was allocated for research funding during fiscal years 1973 and 1974, with an additional fund of $180,000,000 authorized for demonstration grants for new and improved techniques of water pollution control. Additionally, Title II of the 1972 Amendments authorized grants to states, municipalities, and regional agencies for the construction of water treatment and wastewater treatment plants. President Nixon vetoed the amendments when predictions were made that expenditures of over $18,000,000,000 in federal funds would be caused by the construction grant provision of Title II; Congress overrode the President's veto. Nevertheless, the President acted to empound funds allocated for construction grants on the grounds that empoundment was an effective way to reduce inflation. The Supreme Court ruled in February, 1975 that President Nixon had exceeded his authority in this matter; release of the funds was ordered.

Title III of the 1972 Amendments provided for the establishment and enforcement of national water quality standards. Effluent discharge limitations for point sources other than publicly-owned treatment works were to be established by the EPA no later than 1977. These standards are to reflect the best practicable control technology available at that time. The EPA was directed to establish effluent standards by 1983 to eliminate discharge of all pollutants into the nation's waters through application of the then *best available* pollution control technology. Also under Title III, federal enforcement powers were granted. Violation of the 1972 Amendments is punishable by a fine of not less than $2,500 nor more than $25,000 per day of violation and/or imprisonment for one year.

An important facet of the 1972 Amendments was contained in Title IV, which provided for the issuance of permits and licenses for discharge of pollutants under the National Pollutant Discharge Elimination System. Under provision of this system, it is now illegal in the United States to discharge any pollutant from any point source, including publicly-owned facilities, unless the discharge is allowed by a permit granted by the Federal Government. In essence, the issuance of a permit certifies that the discharge complies with current applicable federal water quality standards.

Finally, Title V of the Amendments provides that any citizen who is, or who may be, adversely affected by water pollution may initiate a civil action against the polluter. Under provisions of Title

V, Federal agencies and other offices of the United States Government are subject to lawsuits under charges of alleged water pollution, in the same way that private industry or other governmental agencies are liable. The possibility of citizen suits and class actions for environmental protection is discussed in a later section of this chapter.

In general, the local and state water pollution control activities in the United States were superseded by the federal authorities defined in the 1972 Amendments. At the present time, the individual state governments are responsible for developing state emission controls and regulations which comply with these overall federal control programs. State governments may issue waste discharge permits; however, such permits must comply with the provisions of the 1972 Amendments. The exercise of local and state authority, therefore, has been superseded by federal authority, and such state and local water pollution control activities are of only academic interest. In fact, such activities may constitute a waste of taxpayer money.

The 1972 Amendments to the original Water Pollution Control laws represent the best legislative attempt, to date, at water pollution control in the United States. No immediate changes or further amendments of the Water Pollution Control Act are foreseen (Ref. 2-10).

2-3.4   Solid Waste Management

With the passage of Public Law 89-272, the Solid Waste Disposal Act of 1965, the Federal Government entered into the field of solid waste management. Prior to this time, refuse management had been the province almost exclusively of local governments. The provisions of the 1965 Act called for the demonstration and construction of solid waste management and resource recovery systems to preserve and enchance the quality of air, water, and land resources. The act provided technical and financial assistance to state and local governments and interstate agencies in the planning and development of waste disposal and resource recovery programs. The goals of the Act also included the promotion of national research and development programs to improve solid waste management techniques, to achieve more effective organizational arrangments, to develop new and improved methods of collection, and to establish methods for separation, recovery, and recycling of solid wastes. Emphasis was placed upon development of techniques for environmentally safe disposal of any residues which could not be recycled. Guidelines for all these activities were to be promulgated by the Federal Government. Funds were allocated to provide training grants in occupations involving design, operation, and maintenance of solid waste disposal systems. Originally, the authority to enforce the 1965 Act was delegated to DHEW and the Department of the Interior but, under Reorganization Plan Number 3, these activities were transferred to the Office of Solid Waste Management Programs of the Environmental Protection Agency.

The 1965 Act was amended in 1970 with the passage of the Resource Recovery Act, Public Law 91-512. This Act was designed to transfer responsibility for solid waste management to state and local agencies, in addition to the Federal Government. The 1970 Act directed the EPA to establish guidelines in drawing model statutes for local solid waste management activities. The Act authorized grants for state research and training programs and for the demonstration of projects in resource recovery. Further, in order to emphasize recovery of materials, the Act established a seven-man National Committee on Materials Policy to report to the President on the status of national material resources.

A very important aspect of the 1970 Act was the requirement for the EPA to report on and recommend a system of national disposal sites for the storage and disposal of hazardous substances, including toxic chemical, radioactive, and biological wastes. Work is progressing at the present time in the listing of hazardous materials and in the establishment of proper disposal techniques for such hazardous materials. Included within these activities are the recommendation of treatment methods and stabilization techniques for hazardous materials, the listing of possible disposal sites in given areas for hazardous materials, and the estimate of costs for developing and maintaining the recommended hazardous waste disposal sites.

For the most part, federal solid waste legislation has been impressive on paper but has not functioned well in practice. A primary reason for such failure is that solid waste management, by its very nature, is a localized function. Even in areas of many individual states, comprehensive laws governing solid waste management have not been greatly effective; enforcement has certainly not been effective. Nevertheless, some progress has been made in the establishment of regional solid waste management plans for the states of Colorado, Connecticut, Kentucky, Michigan, New Jersey, Ohio, South Carolina, Tennessee, and Washington. Other states are undertaking similar legislation. However, on the whole, state laws have seldom been enforced or effective.

In summary, solid waste management remains a localized function and must be dealt with on the level of local government with local management activities. However, problems of solid waste management must be assessed in any comprehensive environmental assessment study. Local government capabilities in solid waste management must always be firmly evaluated.

A comprehensive statement of solid waste laws, on a national basis, is contained in references 2-4, 2-30, and 2-31. In addition to the formal laws and provisions noted in these references, a comprehensive treatment of solid waste management activities can be found in references 2-34 and 2-36. A recent indication of some hope for improvement in resource recovery efforts has been voiced by a member of the Senate Interior Committee, Senator Henry M. Jackson (Ref. 2-11). Senator Jackson's statement is typical of many of the statements currently made by individuals concerned with the expenditure of energy in disposal of waste materials and with the vast expenditure of energy in processing new manufactured goods from raw materials. The expenditure of such energy can be greatly decreased through the recycling of waste materials and processing of such secondary materials, rather than the processing of virgin ores. Increasing costs for raw materials should militate against further exploitation of raw materials and should provide incentives for materials and energy recovery efforts directed at solid wastes.

2-3.5 Noise Abatement

Although noise abatement laws were passed in Imperial Rome and in England under the Plantagenets, no comprehensive national legislation for noise control was passed in the United States until 1970. In general, noise abatement laws passed prior to that time were poorly written, technically deficient, and applicable only to small localities. In 1970, with the passage of the Clean Air Act amendments, noise pollution and abatement were included in Title IV of the amendments. Title IV incorporated the Noise Pollution Abatement Act of 1970 and established the Office of Noise Abatement and Control in the EPA. This office was charged with the duty to begin basic research into the generation of environmental noise, the physiological and psychological effects of that noise, and possible measures for abatement of noise. The EPA was assigned the duty of determining techniques for abating noise created by the activities of federal agencies or by other agencies utilizing federal funds. Until the time of the 1970 Act, Federal noise abatement programs were sporadically enforced and had only minimal statutory background. In 1966, a statute had been passed which authorized research into noise created in transportation activities, especially aircraft transportation. In 1968, an amendment to the Federal Aviation Act of 1958 authorized a program to abate and control civilian aircraft noises. In 1970, the passage of the Airport and Airways Development Act enpowered the Federal Aviation Administration to direct airport development policies toward noise abatement and protection of environmental quality near airports. Finally, also in 1970, an amendment to the Federal Aid Highway Act charged the Secretary of Transportation with the duty of minimizing adverse effects from federally-supported highway projects; included in the adverse effects from such projects was the generation of noise during construction as well as during later use of the completed highway.

Federal statutes have been enacted to govern the control and abatement of occupational noise. In general, occupational noise control lies within the province of the Department of Labor under the authority of the Walsh-Healey Public Contracts Act and the Construction Safey Act. The 1970 Williams-Steiger Occupational Safety and Health Act also contains provisions for the protection of workers from occupational noise exposure. However, the most sweeping federal noise pollution control legislation was enacted in 1972 with the passage of the Environmental Noise Control Act as an amendment to Title IV of the Clean Air Act of 1967, and as a new Title V to that act.

The Environmental Noise Control Act of 1972 is based upon the concept that federal action will be required for the effective regulation of noise created by sources which move in interstate commerce. The 1972 Act includes the provisions for the establishment of noise emission standards by the EPA for products manufactured after 1972. Additionally, research on product noise emissions (especially on aircraft noise emissions) and on the effects of noise on human beings and other living organisms was authorized by the 1972 Act. The EPA is required to establish noise criteria similar to the ambient air quality criteria stipulated in the 1970 Amendments to the Clean Air Act. The establishment of these noise criteria was required within nine months after enactment of the 1972 Noise Control Act. The Federal Government is authorized to preempt state and local authority in noise control. Local governments retain the authority to establish noise limits and to regulate the use and operation of products within their jurisdictions, but the federal government can preempt state and local authority when the EPA deems that such authority is not sufficient to prevent noise pollution. The EPA is required to publish reports identifying products considered to be major sources of noise and to regulate the sale of such products. The manufacturers of products identified as major noise sources will be required to guarantee that the manufactured goods will not emit noise at levels higher than those established in the EPA noise emission criteria. Violation of the 1972 Noise Control Act can bring criminal penalties, including fines of $25,000 or one year in prison, for the first violation. In connection with violations, it has been considered a criminal violation to render inactive or to remove any noise suppression devices installed by manufacturers on products covered by the 1972 Act.

The 1972 Act contains provisions directed especially toward the abatement of aircraft noise. Standard methods for measuring aircraft noise are to be established by the EPA, and the EPA is directed to identify critical levels of aircraft noise, below which public health and welfare are guaranteed. The EPA is directed to join the Federal Aviation Administration in determining the degree to which it is technologically feasible to reduce aircraft noise. Furthermore, the 1972 Act authorized the expenditures of up to $22.5 million to state and local noise control agencies over a three-year period in efforts to implement local noise pollution abatement. In general, the control of noise pollution remains a function of local government, and the major mechanism for citizen control of noise pollution is through actions under the common law. The full impact of the 1972 Act has not yet been experienced and, to date, only a few products have been listed as major sources of noise. However, in coming months and years, federal controls on noise pollution are expected to become more comprehensive and stricter.

Other pertinent materials are found in references 2-5, 2-28, and 2-29.

2-3.6   Other Statutes Affecting Environmental Protection

Included in the body of statutory law in the United States, which has an important effect on maintenance and improvement of environmental quality, are certain other statutes not specifically directed toward environmental protection. For example, the Tax Reform Act of 1969 has been utilized to encourage private industry to combat industrial pollution through the provision of favorable amortization procedures for pollution control equipment (Ref. 2-8). Under Section 169 of the Tax Reform Act, a taxpayer is allowed to recover the cost of pollution control facilities over a 5-year period, instead of over the longer period provided for other types of equipment in Section 167 of this Act. The 5-year amortization period is limited to those facilities with a useful life of no more than 15 years, or to that fraction of the equipment costs which can be allocated to the first 15 years of life of the facility. To obtain this benefit, the taxpayer must have state authorities certify that the pollution control facility is in conformity with state programs established under provisions of the Federal Water Pollution Control Act and the Clean Air Act; additionally, the pollution control facility must be certified by a regional administrator of the EPA. Many of the other laws previously cited contain provisions for financial aid for the planning and construction of pollution control facilities. For example, the Federal Water Pollution Control Act Amendments contain provisions for financial aid for demonstration grants. The State And Local Fiscal Act of 1972 (The Revenue Sharing Act) provided allocations, to each state, of a part of the revenues collected by the Federal Government on the basis of specific formulas. The funds so allocated may be used for seven listed expenditures one of which is the control of environmental pollution.

In addition to this type of legislation, a significant impact on environmental protection is anticipated through the passage of some type of Land-Use Planning Legislation in future sessions of Congress (Refs. 2-6, 2-9). Although federal land-use planning laws are not in effect in a comprehensive fashion at the present time, it is inevitable that a national land-use policy will be developed and a land-use planning law will be enacted in the near future. Such laws will have tremendous significance for the conservation and preservation of natural resources.

2-3.7   Summary on Statutory Law

The most comprehensive piece of statutory law concerning environmental protection passed, to date, in the United States is the National Environmental Policy Act (NEPA) of 1969, to be described in following pages. Additionally, a comprehensive body of laws has been passed for the maintenance and enhancement of air and water quality. Federal legislation has been passed to assist in solid waste management activities and to facilitate resource recovery. Federal noise abatement laws also have been enacted. Tax benefits have been granted through passage of the Tax Reform Act of 1969 for the purpose of creating incentives for pollution control activities. Finally, impending passage of land-use legislation and the development of a national land-use policy should have significant effects on environmental protection activities in the United States.

In evaluating the importance of statutory law for environmental protection in the United States, it must be remembered that while federal law preempts state and local ordinances and statutes in many areas, the local and state courts remain a primary source of relief from noise pollution and solid waste problems. The entire body of statutory law, including Public Laws, state laws, and local regulations, forms a background for the assessment of environmental quality. In many cases, environmental quality, with respect to water and air environments, is established through development of water quality and air quality standards by the EPA. Therefore, this body of statutory law forms a necessary background to any assessment activity, in that criteria have been established, defining the quality of the natural environment. These criteria give a baseline against which projected changes in environmental quality may be measured.

2-4. ENFORCEMENT OF ENVIRONMENTAL PROTECTION STATUTES

Because of the basic organization of the government in the United States, the duty to enforce federal legislation has devolved on the Executive Branch. Thus, enforcement of laws aimed at preventing environmental degradation rests with the many offices and agencies within the Executive Branch. Much of the planning for environmental utilization and preservation is carried out by federal agencies. Some of these agencies have come into being as a result of specific enabling legislation; others have grown out of a reinterpretation and restructuring of existing statutes which created other agencies. In general, the agencies within the federal government can be divided into those agencies within the Executive Office of the President, those Departments within the Executive Branch, and other agencies which have received essentially independent charters.

2-4.1 Agencies Within the Executive Office

Within the Executive Office of the President, the foremost agency empowered to curtail environmental pollution is the Environmental Protection Agency, which was created under Reorganization Plan Number 3, in 1970. This plan was an attempt to organize an integrated governmental effort to control and prevent environmental pollution. According to this reorganization plan, bureaus from four then-existing federal agencies and one inter-agency council were combined to form the resulting EPA. Components of the EPA were drawn from the Federal Water Quality Administration, the National Air Pollution Control Administration, the Environmental Control Administration (three bureaus), and the Pesticides Research and Standards Setting Program of the FDA. Additionally, the Pesticides Registration Authority of the Department of Agriculture was included in the EPA.

The EPA began operations as an independent body on December 2, 1970. Its primary duty was to set and enforce standards of environmental quality and to support and encourage research and monitoring in all phases of environmental control and assessment. The EPA is also responsible for furnishing technical and financial assistance to other federal agencies and to state and local bodies. Finally, the EPA has the duty to develop a corps of personnel within the United States to undertake environmental research, monitoring, and improvement.

Other agencies within the Executive Office include the Council on Environmental Quality, the Office of Management and the Budget, and the Office of Science and Technology. The Council on Environmental Quality will be discussed in greater detail in a subsequent section dealing with the National Environmental Policy Act of 1969. The Office of Management and the Budget is assigned the duty of allocating federal funds to governmental agencies charged with the administration of national policy. Thus, indirectly, OMB could exert significant control over pollution abatement activities. To the present time, however, OMB has not assumed a leadership role in this way. Other offices can be established within the Executive Office of the President, at the discretion of the Chief Executive, whenever he deems that technical expertise and advice are required. In this context, such experts often can, and do, significantly influence environmental protection activities in the United States, through the advice they give the President.

2-4.2   Departments in the Executive Branch

A number of departments, intimately associated with the monitoring and preservation of environmental quality, exist within the Executive Branch of the Government. The Department of Agriculture (USDA) is involved through many of its subordinate units in activities which have a profound effect on the environment. Because of the importance of these activities to environmental quality, the Department of Agriculture has been both praised and criticized by environmental groups. For example, the financing of soil and water conservation projects through the Farmers Home Administration of the USDA has drawn praise as well as criticism. Some of these projects have been seen by certain groups as endangering given segments of the environment. In other cases, financing for similar projects has been praised as significant actions to preserve and enhance environmental quality. Other agencies within the USDA include the United States Forest Service, which manages and protects the National Forests of the United States; the National Soil Conservation Service, which provides information on soil management and conservation; the Economic Research Service; the Agriculture Research Service; and the Agricultural Stabilization and Conservation Service; all of which support research in agriculture, economics, plant science, and soil conservation.

Also within the Executive Branch, the Department of the Interior has vast authority to govern the management of natural resources. The Department of the Interior controls the utilization of over 500 million acres of federal land. For example, within the Department of the Interior, the Bureau of Mines engages in research on the utilization of, and exploration for, mineral resources, and maintains data banks on mineral and fuel supplies in the United States. The Bureau of Land Management, in the Department of the Interior, is responsible for direct management of 60% of the total land area of the United States. The Bureau of Reclamation is responsible for the regulation, utilization, and conservation of vast resources of land and water in the southwestern United States. One of the principal activities of the Bureau of Reclamation is the administration of nine hydroelectric plants and the operation of six other hydroelectric plants owned by other government agencies. The United States Geological Survey classifies federal lands with respect to their geological characteristics and the availability of minerals for exploitation. The USGS furnishes countless numbers of maps and circulars of great value to planners and ecologists who are developing environmental assessments and land-use plans. The National Park Service administers 230 field areas and also administers the National Environmental Education Program to provide primary and secondary school curricula with integrated environmental teaching materials. The Office of Water Resources Management sponsors programs devoted to the conservation and enhancement of the nation's water resources. The Bureau of Sports Fisheries and Wildlife engages in widespread environmental conservation and coordination programs and develops comprehensive assessments of river basins. Several other offices contained within the Department of the Interior exercise considerable influence on environmental protection activities in the United States.

The Department of Defense and other departments at Cabinet level within the Executive Branch also have influence on environmental protection activities. In the Department of Commerce, the National Oceanic and Atmospheric Administration, and the Assistant Secretaries for Environment and Urban Systems and for Systems Development and Technology, in the Department of Transportation, exercise considerable influence on environmental quality. In the Department of Defense itself, the United States Army Corps of Engineers has been in the forefront of environmental controversy for years. The Corps has drawn considerable criticism because of its management of navigable waterways and its efforts to furnish flood control and protection along the nation's streams. Because of their widespread occurrence and because of their overall importance, the activities of the Corps must be thoroughly assessed, and rash judgments, either pro or con, must not be made without full evaluation of all available information. Since the passage of NEPA, the Corps has been one of the most successful federal agencies in efforts to adequately assess the environmental effects of proposed activities.

2-4.3 "Independent" Agencies

In addition to the foregoing agencies within the Executive Office of the President and the Executive Branch, several other agencies have been created within the Executive arm of the government which, in essence, have been given statutory authority as independent agencies. Formerly, the Atomic Energy Commission was responsible for granting licenses to individuals or companies for the production of power through nuclear fission and nuclear fusion. Through this activity, the AEC was intimately involved in environmental assessment for a large number of years. Likewise, the Federal Power Commission has been charged with establishing priorities for the generation and transmission of power, and thus it indirectly regulates the exploitation of fossil fuel resources in the United States. In addition to controlling exploitation, the FPC also has had the important task of conserving fossil fuels. Other important commissions within the Executive Branch of the government include the Appalachian Regional Commission, the Federal Maritime Commission, the National Science Foundation, the Delaware River Basin Commission, the General Services Administration, the Tennessee Valley Authority, and the newly created Federal Energy Administration. The last named agency is likely to emerge as a dominant force in regulating attempts to produce power and utilize fuel resources in the United States in such a way as to minimize the detrimental environmental effects of such energy production.

2-4.4 Administrative Law

The activities of the various agencies mentioned above are governed by the rules, regulations and general orders which have been included in the statutes that established the agencies, or in the Executive Orders from the Presidents who have established the agencies. The courts have ruled that the administrative procedures specified by statute and the administrative regulations promulgated by the agencies constitute law. Violation of administrative procedures within an agency can be the grounds for a lawsuit, and several such lawsuits have been filed to obtain control of environmental pollution, on the basis that agencies have not fulfilled the duties imposed on them by the enabling legislation which created those agencies. On the other hand, an administrative agency has authority and jurisdiction only to the extent to which it is granted in the statute or Executive Order which created it. The statutes which have created many of the federal agencies have given those agencies the power to issue cease-and-desist orders, and have also given those agencies the power to take affirmative action in preventing environmental degradation. Therefore, many lawsuits have been brought seeking environmental protection, on the grounds that an agency has not fulfilled its charge as contained in the statutes which created that agency. This approach has met with only limited success because the courts have held that a person claiming injury must exhaust all administrative remedies before obtaining a judicial review of an administrative action or a lack of such action. The courts will not act unless it can be shown that a true controversy exists, and even in truly controversial cases, the courts have not reviewed the facts in a given case but have limited themselves to questions of law in determining whether a given agency has followed the procedures set out in the statues which created that agency. However, considerable ambiguity still exists in the judicial review of administrative actions. In some cases, courts have intervened to either cause or prevent administrative actions; in other cases, the courts have declined to interfere in activities of the Executive Branch of the government.

2-4.5   Summary on Enforcement

A large number of federal agencies are entrusted with the duties to enforce environmental protection measures in the United States and to monitor and improve environmental quality in this country. It is important for persons engaged in environmental assessment and conservation activities to understand the functions of the various federal agencies, since many of these agencies are charged with determining criteria for the measurement of environmental quality and the regulation of discharges and emissions of environmental pollutants. Additionally, the policies promulgated by these agencies have very significant effects on pollution control activities in the United States. Therefore, it is very important to understand the jurisdictions and duties of the various federal agencies, and to closely monitor changes in administrative policy within the agencies of the Executive Branch of the government.

2-5. CONSTITUTIONAL GUARANTEES TO ENVIRONMENTAL PROTECTION

During recent years, various claims have been advanced to the effect that a clean and unsullied environment is guaranteed to every citizen in the United States on the basis of provisions within the Constitution. This attempt to obtain environmental protection through reversion to Constitutional rights and guarantees has met with only limited success.

2-5.1 The Fifth, Ninth, and Fourteenth Amendments

The basic claim for environmental protection supported by constitutional guarantees is that one of the rights promised by the Constitution is the protection of natural resources or the guarantee of an environment free of contamination. In this connection, the Ninth Amendment has been cited. The Ninth Amendment states, "The enumeration in the Constitution of certain rights, shall not be construed to deny or disparage others retained by the people." In addition, the Fifth and Fourteenth Amendments to the Constitution guarantee due process of law for every citizen. The Fifth Amendment states, "No person shall . . . be deprived of life, liberty, or property, without due process of law; nor shall private property be taken for public use, without just compensation." The Fourteenth Amendment states, "No state shall make or enforce any law which shall abridge the privileges or immunities of citizens of the United States; nor shall any state deprive any person of life, liberty, or property without due process of law; nor deny to any person within its jurisdiction the equal protection of the law." In other words, the Fifth Amendment is applicable to the federal government and the Fourteenth Amendment is applicable to state governments. These provisions in the Amendments to the Constitution have been utilized in attempts to guarantee a clean and unsullied environment. The attempts have met with mixed success.

2-5.2   Court Decisions

Several cases have been decided in support of the claims to a constitutional guarantee for a clean environment. In the case of *Hageborn vs. Union Carbide* (5 ERC 1757), however, the court rejected claims to constitutional guarantees against air pollution. In this case, the plaintiffs had complained that air pollution from a Union Carbide plant had deprived them of the right to breathe clean air and to live in a decent environment as guaranteed by the Ninth Amendment, and had deprived them of the right to life, health, and property as guaranteed by the Fifth Amendment. The court opinion, however, cited a number of other decisions as precedents to reject these claims to constitutional rights to a clean environment. The principal case cited was that of *Ely vs. Velde* (451 F. 2d 1130), in which the Fourth Circuit Court declined to give constitutional status to the claims for a right to an undegraded environment. Although a growing number of legal commentators have argued that constitutional protection for the environment is guaranteed, other decisions have rejected this view and the trend at the present time in environmental law is toward a denial of constitutional support for the right to a clean environment. In the case of *Environmental Defense Fund vs. Corps of Engineers* (325 F. Supp. 728, 2 ERC 1260), the claim was again made on the basis of the Fifth, Ninth, and Fourteenth Amendments that the rights of individuals to a clean and undegraded environment would be impaired by a proposed damming of an Arkansas river by the Corps of Engineers. In this case, the court declared that it was not within the province of any Circuit Court to guarantee constitutional protection to the environment. Finally, in the case of *Lindsey vs. Normet* (405 US 56), the United States Supreme Court observed that the Constitution does not provide judicial remedies for all social and economic ills, and rejected a claim to constitutional protection for the environment. These cited court decisions clearly indicate that currently, the Judicial Branch of government does not consider that a constitutional basis exists for claims to environmental protection. However, there are guarantees to due process under the law which influence the question of standing and class actions.

2-5.3 Standing and Class Actions

Under provisions of the Constitution, every individual has a right to due process of law. However, this does not automatically give him standing in court. In other words, he may not have the right to file a lawsuit and obtain judicial resolution of a controversy. To show standing, a person must be able to prove that he has a significant interest in a particular controversy and that he will be affected in some way by the decision given by the court in that particular controversy. One of the leading cases on standing with respect to environmental matters has been the Supreme Court decision in the case of *Sierra Club vs. Morton* (405 US 345). In this case, the conservation group sought the assistance of the court in preventing the development of the Mineral King Valley for recreational purposes. The court gave the important opinion that environmental well-being, like economic well-being, could be the basis for the claim of injury, and that a person suffering injury to his environmental well-being would have standing to sue. However, in this particular case, since the Sierra Club had not alleged that any of its members would be directly affected by the development, the court denied the pleas of the Sierra Club to prevent the recreational development of the valley. Usually, it is a relatively simple matter to show that individuals would be affected in a particular development. Recent court decisions indicate that if it is possible to show injury to scenery, historic objects, natural materials, or wildlife, then this injury is a sufficient basis for a lawsuit by a person or group of persons who will feel that injury in a direct way. Citizens are specifically given standing and access to federal courts through provisions of the Clean Air Act, the Federal Water Pollution Control Act, and the Noise Control Act of 1972. These court rulings bring up the topic of class actions, which needs some clarification.

In a class action, one or more individuals represent themselves as characteristic of a class of persons who may be injured or deprived in some way, as a result of an existing or a proposed action. In general, class action suits have been held valid only under certain circumstances. A class action has been allowed only when the class of persons cited is so numerous that a gathering of all members of that class cannot be attained. Additionally, class actions have been allowed only when it is possible to show that the particular question of law or fact in the case pertains to all members of that class. The claims of the person(s) purporting to be representative of a class must be typical of the claims or defenses of the class itself. A final test has been applied in that it is held that class actions cannot be considered valid if the parties allegedly representing a class can be shown to be unfair or inadequate representatives of that class. Within the context of these restrictions, however, redress for environmental degradation can be obtained in federal and local courts through class actions.

2-5.4   Summary on Constitutional Claims

Several attempts have been made to obtain redress for environmental degradation on the basis of constitutional claims resting on the language embodied in the Fifth, Ninth, and Fourteenth Amendments to the Constitution. However, for the most part, the courts have denied the constitutional basis of such claims. On the other hand, a clear constitutional right to due process does exist. Through the establishment of standing, on the basis of legal precedent or on the basis of statutory guarantees embodied in new environmental legislation, plaintiffs may prevent further degradation of the environment and, in some cases, may obtain redress for past damages. Class actions may be undertaken to stop environmental degradation only if the persons bringing the action can be shown to be truly representative of the class of persons who would be injured by the given action. Within this framework of restrictions and guarantees, it is possible for individuals to obtain judicial review of private and administrative activities, with respect to environmental protection and pollution.

2-6. SUMMARY

2-6.1 Common Law

In attempts to achieve environmental protection through legal means, the common law in the United States can be utilized in several ways. The three most common forms of action in civil lawsuits under the common law are claims of negligence, nuisance, and trespass. The importance of trespass as a means of obtaining judicial review of actions, or payment for damages suffered, is not as significant as the importance of nuisance and negligence. The passage of recent environmental legislation has made it easier to prove the negligence of polluting individuals and facilities, and negligence actions should increase in number and importance in the future, in efforts to obtain environmental pollution control. Nuisance claims remain a principal way for the individual citizen to obtain relief from noise pollution and from odors resulting from improper solid waste management.

2-6.2 Environmental Protection by Statute

An entire body of statutory law has been enacted by the Congress of the United States in attempts to achieve environmental protection through legislative actions. The most important statutory laws guaranteeing environmental quality in the United States are the 1970 Clean Air Act Amendments, the 1972 Federal Water Pollution Control Act Amendments, the 1972 Noise Pollution Control Act, and the 1970 Resource Recovery Act. These laws not only provide for the establishment of environmental quality criteria by federal agencies, but also enable such agencies to impose fines and jail sentences for violations of pollution prevention codes. In the end, however, the full effectiveness of this body of statutory law will rest with the enforcement agencies within the Federal Government who are charged with the duty to adequately administer these laws.

2-6.3 Administration of Statutory Laws

A large number of agencies exist within the federal government for enforcement of statutory laws intended to guarantee environmental protection. The most important federal agency in this regard is the United States Environmental Protection Agency. This agency is charged with the responsibility to establish air quality and water quality criteria and to eliminate discharges of pollutants into the nation's streams. A thorough appreciation of the procedures and activities of various federal agencies is necessary for any person attempting environmental assessment or environmental protection, since the activities of many federal agencies have considerable importance in the management of our natural resources. The activities of the EPA in establishing environmental criteria serves to give baseline data against which changes in environmental quality can be measured. This is essential for adequate environmental assessment. Finally, administrative policy reflects trends in the management of resources, as conducted by the Executive Branch of the Government. These trends are valuable and significant indicators of the directions taken by the Executive Branch in maintaining and enhancing environmental quality.

2-6.4 Constitutional Claims to a Clean Environment

Claims have been set forth for a guarantee to a clean and unsullied environment, on the basis of the Fifth, Ninth, and Fourteenth Amendments to the Constitution. However, in general, the courts have held in various precedent-setting cases that such constitutional claims are not valid. Nevertheless, the constitutional guarantees to due process of law can be utilized in establishing standing in court and in granting citizens the right to bring lawsuits for environmental protection. These citizen's rights to sue to guarantee environmental quality are also contained within the texts of many of the most recent environmental protection statutes enacted by Congress. Finally, class actions, if properly brought, under certain restrictions, can be a powerful tool for the redress of environmental wrongs. A considerable increase in the number of such class actions should occur in coming years.

REFERENCES

2-1. Arbuckle, J. G., Schroeder, S. W., and Sullivan, T. F. P., *Environmental Law for Non-Lawyers,* Edited by R. A. Young, Government Institutes, Inc., Bethesda, Maryland, 2nd Ed., 1974, pp. 1–45.
2-2. ibid., p. 115.
2-3. ibid., p. 65.
2-4. ibid., p. 191.
2-5. ibid., p. 172.
2-6. ibid., p. 201.
2-7. Hanks, E. H., Tarlock, A. D., and Hanks, J. L., *Environmental Law and Policy,* American Casebook Series, West Publ., St. Paul, Minnesota, 1975.
2-8. Reitz, Arnold W., Jr., *Environmental Law,* North American International Co., Washington, 1972.
2-9. Reitz, Arnold W., Jr., *Environmental Planning: Law of Land & Resources,* North American International Co., Washington, 1974.
2-10. *Conservation Report,* Rep. No. 39, Dec. 13, 1974, National Wildlife Federation, Washington, p. 531.
2-11. *Phoenix Quarterly,* Vol. 6, No. 4, Winter, 1975, Institute of Scrap Iron and Steel, Inc., Washington, p. 3.

Air Quality Laws

2-12. P.L. 84-159. Air Pollution Control Research and Technical Assistance Act of 1955.
2-13. P.L. 85-365. 1959 Extension of P.L. 84-159.
2-14. P.L. 86-493. Study of Motor Vehicle Discharges, 1960.
2-15. P.L. 88-206. Clean Air Act of 1963.
2-16. P.L. 89-272, Title II. Motor Vehicle Air Pollution Control Act of 1965.
2-17. P.L. 89-675. Clean Air Amendments of 1966.
2.18. P.L. 90-148. Air Quality Act of 1967.
2-19. P.L. 91-604. Clean Air Amendments of 1970.

Water Quality Laws

2-20. Rivers and Harbors Appropriations Act of 1899, 30 Stat. 1151.
2-21. P.L. 84-660. Federal Water Pollution Control Act of 1956 (FWPC Act).
2-22. P.L. 87-88. FWPC Act Amendments of 1961.
2-23. P.L. 89-234. Water Quality Act of 1965.
2-24. P.L. 89-753. Clean Water Restoration Act of 1966.
2-25. P.L. 91-224. Water Quality Improvement Act of 1970.
2-26. P.L. 92-500. FWPC Act Amendments of 1972.

Other Environmental Laws

2-27. P.L. 91-190. National Environmental Policy Act of 1969.

2-28. Noise Pollution and Abatement Act of 1970—see Title IV, Clean Air Act Amendments of 1970.

2-29. P.L. 92-574. Noise Control Act of 1972.

2-30. P.L. 89-272. Solid Waste Disposal Act of 1965.

2-31. P.L. 91-512. Resource Recovery Act of 1970.

2-32. P.L. 91-172. Tax Reform Act of 1969.

2-33. P.L. 93-159. Emergency Petroleum Allocation Act of 1973.

2-34. Hagerty, D. J., Pavoni, J. L., and Heer, J. E., Jr., *Solid Waste Management*, Van Nostrand Reinhold Co., New York, 1973, 310 pp.

2-35. P.L. 93-319. Energy Supply and Environmental Coordination Act of 1974.

2-36. Pavoni, J. L., Heer, Jr., J. E., and Hagerty, D. J., *Handbook of Solid Waste Disposal*, Van Nostrand Reinhold Co., New York, 1975, 533 pp.

3

The National Environmental Policy Act (NEPA)

3-1. INTRODUCTION

The National Environmental Policy Act of 1969, Public Law 91-190, was signed into law on January 1, 1970 by President Richard M. Nixon. This law has been described by various sources as the most important and significant piece of environmental legislation ever enacted by the United States Congress. This statement may seem to be debatable by some who recognize that this legislation is, at present, less than eight years old and is certainly still in a developing state, as the courts continue to rule upon its interpretation. Nevertheless, it seems destined to play a role in environmental protection in the United States that will equal or surpass that played by previous legislation such as the Federal Water Pollution Control Act, the Air Quality Act of 1967, the Clean Air Act Amendments of 1970, and other similar laws. While this national environmental policy act was introduced into Congress, passed, and signed into law by the President in a period spanning just a few months, it is important to realize that the basic idea of such a national environmental policy had been discussed in many forms in Congress and in the country as a whole, for more than a decade. Congress had considered the goals of a national environmental policy for many years prior to the actual passage of the act.

A study made in 1970, about the time of the passage of the act, by the Library of Congress, Legislative Reference Service (Ref. 3-1), identified over 58 offices and bureaus in nine departments and over 15 other federal agencies concerned with various aspects of the environment. Lacking in this array of concern for the environment and conservation of resources was any type of a national policy on these issues.

3-2. BACKGROUND OF NEPA AND LEGISLATIVE HISTORY

Late in 1959, Senator James Murray (D-Montana) introduced his Bill S.2549, "The Resources and Conservation Act," which, in his words, called for the Congress to issue a national statement of conservation of resources and an environmental policy, and at the same time, to create a high-level council of environmental advisors. This bill was patterned after his earlier 1946 Employment Act. In this legislation, Senator Murray proposed three ideas, new with respect to the environment, but similar to those in the Employment Act. These were: (1) An Environmental Advisory Council to the President; (2) A joint committee of the houses of Congress on resources and conservation; and (3) An annual report on resources and conservation to be submitted to the Congress by the President. The bill contained one other important feature, a declaration of national policy on conservation of resources. This was the seed from which, a decade later, after much legislative maneuvering, was to grow the National Environmental Policy Act of 1969.

Section 2 of S.2549 stated Senator Murray's version of the policy:

The Congress hereby declares that it is the continuing policy and responsibility of the Federal Government with the assistance and cooperation of industry, agriculture, labor, conservationists, state and local governments, and private property owners, to use all practicable means including coordination and utilization of all its plans, functions, and facilities for the purpose of creating and maintaining in a manner calculated to foster and promote the general welfare, conditions under which there will be conservation, development, and utilization of the natural resources of the Nation to meet human, economic, and national defense requirements, including recreational, wildlife, scenic, and scientific values and the enhancement of the national heritage for future generations.

Public hearings were held on the bill; however, the Eisenhower administration opposed the measure. The Office of Management and the Budget was cool to a new council of advisors in the Executive Office of the President, and other federal agencies did not care for a coordination of their activities in the resources area by another agency. Because of much opposition, the bill died in committee, but the seed had been planted in the legislative process.

3-2.1 Legislation of the Sixties

Additional legislation addressing itself to the problem of resources and conservation was introduced during the 1960's. At various times, legislation addressing itself to the problem of a national environmental policy was introduced by Senators Nelson, Jackson, and Kuchel.

In tracing the historical development of the policy which was finally proclaimed by the passage of NEPA, it is important to note several items of legislation proposed in the Congress; of importance also, was the work of some of its committees. In June 1968, a subcommittee of the House of Representatives Committee on Science and Astronautics published a report entitled "Managing the Environment." This report was widely circulated and was to be the first of several documents after which NEPA was patterned. In this report to the second session of the 90th Congress, the subcommittee, in essence, blamed most of the nation's then existing environmental problems on the fact that governmental efforts in decision making were fragmented and dispersed through an excessive number of governmental agencies, with no single group having either the authority, or the responsibility, to coordinate the nation's efforts in this regard. The report contained several other sections worthy of note for their background to the Environmental Policy Act. It was proposed that the Department of the Interior be designated as the lead agency to coordinate the environmental management operations of all Federal programs. It contained a suggestion for the transfer of many existing agency functions from several departments to the Department of the Interior.

The report further recommended that each Federal agency with responsibility for environmental actions designate a senior official for environmental programs. These top-level officials would then form, together with the chairman of the Federal Council for Science and Technology, an "Environmental Cabinet." Still another suggestion was that:

> The best means of gaining a long-term rational management is to generate an informational base and provide a policy to all operational programs which will cause individual decision-makers to act in harmony with the entire system (Ref. 3-2).

This familiar theme was to recur over and over in legislative proposals and be, in essence, the idea finally incorporated in the action-forcing procedures of the 1969 act. Some months earlier, in June 1967, this same theme was expounded in a report from the Department of Health, Education, and Welfare entitled "Strategy for a Liveable Environment" (Ref. 3-3). One section of this report stated that what the country urgently needs . . .

> . . . is an overview of the entire question of environmental health and its interrelated components, not only water pollution, air pollution, solid wastes, but, also, noise, crowding, radiation, traffic safety, and ailments which can be related to these factors.
>
> The American public must develop this overview, a sensitivity to the scope and limitations of the environment.

3-3. DEVELOPMENT OF AN ENVIRONMENTAL POLICY

In mid-1968, the Senate Committee on Interior and Insular Affairs and the House Committee on Science and Astronautics joined in sponsoring a Senate-House Colloquium to discuss the possibility of Congress enacting a national policy on the environment. Prior to this meeting, the Senate committee received a special report entitled "A National Policy for the Environment" (Ref. 3-4). This was another in the list of reports that were to become important documents in the later development of NEPA. One result of the joint committee colloquium was the publication of a "Congressional White Paper on a National Policy for the Environment" (Ref. 3-5). This so-called White Paper contained much that was later enacted as Section 101 of the National Environmental Policy Act.

3-4. LEGISLATIVE DEVELOPMENT OF NEPA

The actual immediate legislative history of NEPA is usually traced to two bills: H.R. 6750, introduced February 17, 1969, and Senate Bill 1075, introduced the following day (by Senator Jackson). It should be noted that both these bills were introduced just ten and a half months prior to the final enactment of the bill as it exists today. Senator Jackson's bill S.1075, if enacted into law unamended, would have made the Secretary of the Interior solely responsible for the coordination of national environmental research, and at the same time would have created what is now known as the Council on Environmental Quality (CEQ). Congressman Dingell's bill (H.R. 6750) also called for the creation of a commission or council on environmental quality. It is interesting to note that Congressman Dingell's bill was introduced as an amendment to the Fish and Wildlife Coordination Act.

It has been speculated that both Senator Jackson and Congressman Dingell purposely omitted, from their respective bills, a detailed statement of a national environmental policy, in order that the bills, when introduced into the Congress, would be referred to their own respective committees, wherein they could then be amended if so desired.

In some respects, both of the bills introduced into Congress were modeled after the previously referred to Employment Act of 1946 (Public Law 304). This Act had created a three-man Council of Economic Advisors, whose duty it is to advise the President in matters pertaining to the government's actions and responsibilities, with respect to the national economy. It should be noted that both bills, as well as the finally enacted law, called for the creation of a similar three-man council; in this instance, a council of advisors on environmental quality. The framers of the legislation apparently were pleased with the work of the Council of Economic Advisors and their easy access to the Executive Office of the President; the senators apparently decided to pattern the Council of Environmental Advisors along similar lines.

The council proposed in the Dingell bill had four duties assigned to it in the text of the law:

1. To gather timely and authoritative information concerning the conditions and trends in environmental quality;
2. To appraise the resources programs and policies of the Federal Government;
3. To develop and to recommend to the President national policies on environmental affairs; and
4. To assist the President in the preparation of his annual state of the environment report to Congress.

3-4.1   Action in the Senate

Senator Jackson's committee conducted a hearing on his bill (S.1075) in April of 1969 (Ref. 3-6). Several weeks later, Senator Jackson introduced an amendment to the bill, which was to be "A Declaration of National Environmental Policy." This amendment stated that "each person has a fundamental and unalienable right to a healthful environment." A second provision of the same amendment provided "for the determination by a responsible federal official of the environmental impact of any proposed federal agency action." While this latter wording did not remain in the act as it was finally introduced, it obviously is the predecessor of the now well-known section 102(2)C requirement for an Environmental Impact Statement. This apparently was the first time that Congress had placed the responsibility for determining the environmental effect of their actions, on those federal agencies which, in many cases, previously had contended that they did not have judicial authority to consider such impacts.

3-4.2 Action in the House

Congressman Dingell's bill took a somewhat different route through the legislature, with hearings being held for seven days in the summer of 1969. Much of the discussion at the hearings concerned the need for the Council of Environmental Advisors in the President's Office. Many witnesses (Ref. 3-7) felt that such a council would be a needless duplication of the Office of Science and Technology. Others debated the size of such a council, with recommendations of three to eight members being most frequently mentioned. The House Committee on Merchant Marine and Fisheries reported out the Dingell bill favorably in July, as H.R. 12549.

In order for the sponsors to obtain a favorable ruling on the bill from the House Rules Committee, it was necessary for them to accept several amendments which, at the time, were not particularly welcome. Among the amendments attached to the bill was one calling for the evaluation of *all* environmental impacts of an action, rather than restricting the study to the impacts on fish and wildlife. The second amendment stated, "Nothing in this act shall increase, decrease, or change any responsibility of any federal official or agency" (Ref. 3-8). Had this latter amendment not been dropped in the subsequent Senate-House Committee Conference on the bill, it, in effect, would have nullified the requirement of an environmental impact statement by most federal agencies. The third amendment provided specific monetary authorizations for the proposed Council on Environmental Quality. This provision was added with the remark that the Council itself was unnecessary, ". . . but in any event, there was no need to give it a blank check." The amended bill was passed by the House in September, 1969.

At the earlier hearings on S1075, Senator Jackson had insisted that the Council membership remain at three, as introduced in his bill. At the same time, at the insistence of several witnesses and to gain additional support for his bill, Senator Jackson introduced an amendment to his bill which formally stated a national environmental policy. The amendments, intended to become Title I of S.1075, state that in undertaking any actions, federal agencies should be aware of the environmental impact of those actions, so that the nation might:

(1) fulfill the responsibilities of each generation as trustees of the environment for succeeding generations;

(2) assure for all Americans safe, healthful, productive, and esthetically and culturally pleasing surroundings;

(3) attain the widest range of beneficial uses of the environment without degradation, risk to health or safety, or other unintended, unanticipated, and undesirable consequences;

(4) preserve important historic, cultural, and natural aspects of our national heritage, and maintain, wherever possible, diversity and variety;

(5) achieve a balance between population and resource use which will permit high standards of living and a wide sharing of life's amenities; and

(6) enhance the quality of renewable resources and approach the maximum attainable recycling of depletable resources (Ref. 3-9).

This same amendment directed that all laws, policies, and regulations of the Federal government be administered in accordance with the environmental policy set forth in NEPA. Additionally, it enumerated the requirements of an interdisciplinary approach to planning, an identification of means to quantify environmental values, and the preparation of an Environmental Impact Statement. Much of the substance of this amendment found its way, unchanged, into the enacted final bill. The amendment spelled out in detail, as Section 102(3), the specifics of the required statement (Ref. 3-10). It stated that agencies must:

(3) include in every recommendation or report on proposals for legislation or other significant Federal actions affecting the quality of the human environment, a finding by the responsible official that:

(i) the environmental impact of the proposed action has been studied and considered;

(ii) any adverse environmental effects which cannot be avoided by following reasonable alternatives are justified by stated considerations of national policy;

(iii) local short-term uses of man's environment are consistent with maintaining and enhancing long-term productivity; and

(iv) any irreversible and irretrievable commitments of resources are warranted.

This is the action-enforcing provision which, from the outset, had been envisioned by Senator Jackson as essential to the effectiveness of any national environmental policy.

On July 9, 1969, the Senate Interior Committee ordered the amended version of S.1075 reported to the full Senate for consideration. The amended bill was drastically different from the bill introduced by Senator Jackson. The bill contained: as Title I, the National Environmental Policy Statement; as Title II, an Ecological Research Program; and as Title III, a Board of Environmental Quality Advisors instead of a Council. In addition, the title of the bill was changed to "The National Environmental Policy Act of 1969." On July 10, under an informal procedure normally used to dispense with uncontested legislation, the Senate, after several parlimentary maneuvers, passed S.1075, amended, and with its title changed (Ref. 3-11).

Thus, after some five months of legislative hearings, meetings, compromises, and objections from the administration, the National Environmental Policy Act passed the Senate. There was no debate, and the whole procedure, witnessed by fewer than seven senators, took less than five minutes to accomplish. The Senate could not have been aware of the far-reaching effects of the "action-forcing" provisions of the bill as passed.

3-4.3 Congressional Compromise

The House of Representatives passed Representative Dingell's bill on September 23, 1969, and requested a conference in order to compromise with the Senate on the differences in the two bills, as

Fig. 3-1   Major dam under construction on the Ohio River (*courtesy* Louisville District Corps of Engineers, Department of the Army). Comprehensive environmental planning and assessment will prevent the construction of ill-conceived public works or lead to mitigation of detrimental effects through suggested changes in the design or operation of such facilities.

enacted, before the House sought final Congressional passage of the bill.

In the Senate debate before a conference committee, two more important compromises by Senator Jackson were required to obtain the support of Senator Muskie, then chairman of the Senate Public Works Committee, who had objected to the rapid passage of S.1075 without time for his committee to hold hearings on the bill. In an agreement between the senators, Section 102 was amended to require that the Environmental Impact Statement be made public and be reviewed by agencies with special expertise or responsibility for environmental protection. This requirement was added as follows (Ref. 3-12):

Prior to making any detailed statement, the responsible Federal Official shall consult with and obtain the comments of any established agency which has jurisdiction by law or special expertise with respect to any environmental impact involved. Copies of such statement and the comments and views of the appropriate Federal, state, and local agencies, including those authorized to develop and enforce environmental standards, shall be made available to the President, the Board of Environmental Advisers, and to the public as provided by 5 U.S.C. 552 and shall accompany the proposal through the existing agency review processes.

The second part of the compromise resulted in a new Section 103 of the bill. This new section stated (Ref. 3-13):

Nothing in section 102 shall in any way affect the specific statutory obligations of any Federal agency (a) to comply with criteria or standards of environmental quality, (b) to coordinate or consult with any other Federal or State agency, or (c) to act, or refrain from acting contingent upon the recommendations or certification of any other Federal or State agency.

With these major changes, and several other minor ones, the Senate agreed to send the bill to a conference committee with hardly any debate on the bill as altered by the private compromise. Some observers have interpreted this absence of debate as a sign of lack of interest, by most senators, in environmental legislation at that time.

3-4.4 Conference Committee, Compromise, and Passage

In the conference committee action, several additional substantial changes were made. One notable amendment accepted by the committee was a provision that the air and water standards set by previous legislation would not be affected by the new act. A second change of great significance was the altering of the wording from a requirement for "finding" the environmental impact, to a requirement for a "detailed statement" of the impact; for the first time, a new provision for the description of all alternatives to the proposed action was included. At this stage, the Senate compromise requirement was also retained (that the designated responsible federal official be required to consult with, and obtain comments from, other affected federal agencies before preparing the detailed impact statement). Thus, the requirement of Section 102(2)C for the environmental impact statement, as we know it today, was beginning to take shape, bit by bit.

Any Congressional conference committee meeting is a bargaining session. The House and the Senate committee members must each "give" a little in order to develop a bill upon which they can agree. In this particular committee, the principal compromises were the acceptance by the House of the Senate version of Title I, in exchange for Senate acceptance of the House version of the Environmental Council.

As the Senate-House Conference Committee continued its work, additional changes took place before the bill took its final form. Another important inclusion as a part of Section 102 was the requirement that the agencies comply with the action-forcing provisions "to the fullest extent possible."

Another point of some disagreement was a statement in Title I of the bill, which in Sec. 101(c) stated, " . . . each person has a fundamental and inalienable right to a healthful environment." It was contended by Representative Aspinall that, if so stated, this section would " . . . be seized upon as the basis for numerous lawsuits." This was later changed to read, ". . . each person should enjoy a healthful environment and that each person has a responsibility to contribute to the preservation and enhancement of the environment."

When the conference report was brought to the Senate floor later, Senator Jackson spoke against the change; he stated (Ref. 3-14):

> I opposed this change in conference committee because it is my belief that the language of the Senate passed bill reaffirmed what is already the law of this land; namely, that every person does have a fundamental and an inalienable right to a healthful environment. If this is not the law of this land, if an individual in this great country of ours cannot at the

present time protect his right and the right of his family to a healthful environment, then it is my view that some fundamental changes are in order.

To dispel any doubts about the existence of this right, I intend to introduce an amendment to the National Environmental Policy Act of 1969 as soon as it is signed by the President. This amendment will propose a detailed congressional declaration of a statutory bill of environmental right.

To date he has not introduced such an amendment to NEPA as it was finally enacted.

The conferees agreed upon a bill which was accepted by the Senate, after some debate, on December 20, 1969. On December 22, the conference report was called up in the House. The major opposition was from Representative William Harsha, who almost prophetically stated (Ref. 3-15):

The impact of S.1075, if it becomes law, I am convinced would be so wide sweeping as to involve every branch of the Government, every committee of Congress, every agency, and program of the Nation. This is such an important matter that I am convinced that we here should consider it very, very carefully and make a clear record as to exactly the direction in which we wish the various elements of our Government to move.

I regret that so important a matter is being handled in so light a manner. I realize the Members desire to adjourn for Christmas and that the hour is late and that we are all tired, but this is no subject to merely brush aside. I had hoped that this matter could be laid over until Congress reconvenes, providing the Congress with ample time to fully understand the complete ramifications of this legislation.

However, the House approved the report, and the National Environmental Policy Act was passed and sent to the President for his signature.

The President signed the legislation on January 1, 1970 as Public Law 91-190. The National Environmental Policy Act thus became the first new law of the decade of the seventies. In the final rush to passage, the bill, because of Congressional procedures, was not open to amendments from the floor. This circumstance has caused some persons to complain that this most important bill was rushed into law without due consideraton by Congress as a whole, although it had taken almost ten years to enact NEPA. The validity of this opinion seems to be supported, as will be described in detail later in this chapter, by many diverse and far-reaching court decisions. These decisions, without a doubt, have extended the bill well beyond the limits of what many Congressmen perhaps intended. While there has been no attempt made to amend NEPA to date, a study of court decisions relative to NEPA, made over the last five years shows that it has been changed greatly from the law drafted by the joint conference committee in late 1969.

3-5. NATIONAL ENVIRONMENTAL POLICY ACT OF 1969

The entire statement of the act itself is reproduced as an appendix to this text. However, in view of the fact that a multiplicity of court rulings, with various interpretations of many sections of the Act, have been given to date, it is perhaps wise to analyze the Act section by section. In order to best appreciate the requirements of the Act, the reader should look closely at what actually was required by the bill at the time of its passage, and then look at what is currently required by the law as a result of the various court interpretations.

3-5.1 Purpose of the Act

The purpose of the act, as stated in Section (2) is ". . . to declare a national policy which will encourage a productive and enjoyable harmony between man and his environment; to promote efforts which will prevent or eliminate damage to the environment and biosphere and stimulate the health and welfare of man; to enrich the understanding of the ecological systems and natural resources important to the nation; and to establish a council on environmental quality."

Section 101 is the declaration of a national environmental policy, as well as a declaration of national goals for implementing that policy. In remarks made earlier before Congress, Senator Jackson related his feelings in regard to a national environmental policy by saying,

> A statement of environmental policy is more than a statement of what we believe as a people and as a nation. It establishes priorities and gives expression to our national goals and aspirations. It provides a statutory foundation to which administrators may refer for guidance in making decisions which find environmental values in conflict with other values.
>
> What is involved is a Congressional declaration that we do not intend, as a government or as a people, to initiate actions which endanger the continued existence or the health of mankind: that we will not intentionally initiate actions which will do irreparable damage to the air, land, and water which support life on earth (Ref. 3-16).

The same section of the act requires that the federal government should use all practical means to secure a livable environment for succeeding generations and to ensure that all Americans have a safe, healthful, and aesthetically pleasing surrounding. It also requires the preservation of important historical, cultural, and natural aspects of our heritage, as well as requiring a balance between population and the use of resources, which will permit a high standard of living consistent with a livable environment. As pointed out above, Congress, in this section, further recognized that not only should each person enjoy a healthful environment, but also that each person has a responsibility to contribute to the preservation and enhancement of that environment; they stopped short of saying that each person has a Constitutional or God-given right to a healthful environment.

Section 102 is the most important section of the act with respect to the requirement for an environmental impact statement. This section directs that all agencies of the federal government should utilize a systematic, interdisciplinary approach in all planning and decision making. All such agencies must identify and develop methods, guidelines, and procedures in their decision-making that will ensure unqualified evaluation of environmental considerations along with economic and technical considerations. Section 102(2)C is actually the heart of the environmental policy act, as it pertains to environmental impact statements. The requirements and ramifications of this section are described more fully in the following chapters of this text.

Section 102(2)C states, in general terms, that every recommendation or report on proposals for legislation or other major federal actions *significantly affecting* the quality of the human environment shall require a detailed statement by the designated responsible federal official, with respect to the predicted impact of such proposed action.

The five items *specifically required by the act* to be addressed in the detailed statement are: 1) The environmental impact expected of the proposed action; 2) Any adverse environmental effects which cannot be avoided should the proposal be implemented; 3) Alternatives to the proposed action; 4) The relationship between local short-term uses of man's environment and the maintenance and enhancement of long-term productivity; and 5) Any irreversible and irretrievable commitments of resources which would be involved in the proposed action should it be implemented. This section further requires, as described in the previous paragraphs, that the federal official responsible for making the impact statement should consult with representatives of any other agency that has jurisdiction by law or has special expertise with respect to any of the expected impacts. Copies of the impact statement, when prepared, are required to be sent to the appropriate federal, state, and local agencies for comment. NEPA further requires that the public shall be provided with copies of the statement and appropriate comments.

3-5.2   Title II of NEPA

Title II of the act establishes, as previously noted, the Council on Environmental Quality (CEQ). It should be noted that the Council on Environmental Quality, as finally constituted in the law, consists of a three-man committee in the Executive Office of the President. The three members, according to Sec. 202, shall be persons who, as a result of their training, experience, and attainments are exceptionally well qualified to analyze and interpret local and national environmental trends. Their primary duty shall be to assist and advise the President in his annual environmental quality report to Congress, which is required by Sec. 201 of the Act. In conjunction with its analysis of current trends, the Council is further directed to devise any additional procedures which may be required to satisfy the intent of the national environmental policy as declared by Congress. All of these findings and conclusions also are to be presented to the President for use in his annual environmental quality report.

By Executive Order 11514, CEQ was required to prepare a set of guidelines to be followed in environmental planning or decision-making processes that could include legislation or other federal agency actions (Ref. 3-17). CEQ has distributed such guidelines, and they were published in the Federal Register (Ref. 3-18) under the title "Council on Environmental Quality Statements on Proposed Federal Actions Affecting the Environment—Guidelines." The Council's main role, therefore, has been, and continues to be, primarily one of coordination and recommendation, rather than one of statutory restriction.

In one Act, Congress has established the environmental policy goals for the nation, has constituted a three-man advisory council to the President, whose duties include a monitoring of environmental trends and, most important of all, has required a detailed statement on the predicted environmental impact of any proposed Federal action. Unfortunately, in passing the act, the Congressmen, because of their rush to enact the law and because of the brevity of the act itself, left several important loopholes that have since necessitated extensive elaboration and court interpretation of NEPA.

3-6. NEPA IN THE COURTS

As with any federal law, NEPA must be interpreted by the Judiciary, and enforced by the Executive Branch. Although the act was officially signed into law by the President, the mere existence of such a law did not ensure complete compliance by the various Federal Agencies to which the law was directed. In initial Senate hearings, previously mentioned, which ultimately led to several of the important amendments to NEPA, a conversation between Senator Jackson and Dr. Lynton Caldwell, then of the University of Indiana, suggested that probably the Office of Management and Budget (OMB) would develop into the main enforcement agency for the implementation of NEPA. Senator Jackson expressed his views with respect to this matter when he said, "The Office of Management and Budget should exercise prudence and discretion in requiring that environmental policies and standards be adhered to in connection with the responsibilities of the Federal Establishment." Dr. Caldwell felt that OMB ". . . should be authorized and directed to particularly scrutinize administrative action and planning with respect to the impact of legislative proposals and public work projects on the environment." It has been suggested that even the wording ". . . legislative proposal and recommendations" which actually occurs in Sec. 102(2)C of the Act may, in fact, stem from these conversations. It apparently was the idea of Dr. Caldwell and Senator Jackson that the Office of Management and Budget, as controllers of the funds allotted by Congress, could withhold these funds if the various agencies did not comply with the requirements of NEPA, specifically the requirements with respect to the preparation of the required environmental impact statements. However, the Act gives no specific authority, nor does it allot any responsibility in this matter to OMB. As a result, OMB personnel have evaded the assumption of any such power, principally by their failure to create, as requested, specific CEQ-type Guidelines for use in legislative clearance processes by the Office of Management and Budget.

As mentioned previously, the creation of CEQ, by NEPA, did not entail the delegation of any specific regulatory powers to CEQ. In this regard, CEQ is even more handicapped than OMB as a potential regulatory agency to enforce the act as Congress intended, since CEQ lacks even the institutional jurisdiction over funding that OMB has by statute. Thus, to date, CEQ participation has remained at the advisory level of enforcement as outlined in NEPA. The most forceful activities that CEQ has undertaken in this regard have been provoked by cirumstances which prompted the Council to make exceptionally strong recommendations to the President. Several of these recommendations have resulted in executive action by the President to halt some proposed agency activity which, in the opinion of CEQ and various citizen environmental groups, would very strongly and adversely affect the nation's environmental quality.

Another serious weakness in the Act itself is the ambiguity of parts of Section 102. This section makes no provision for final and ultimate disposition of environmental impact statements. NEPA does not delineate specifically where, and to whom, they shall be sent, how long they are to be under review, or what ultimately is to be done with them.

The major reasons for 1) recounting the history of environmental legislation prior to the enactment of NEPA, 2) detailing what is required by the act itself, and 3) describing the lack of acceptance of responsibility for enforcement of the act, is to establish the background for an extensive and comprehensive study of court actions with regard to NEPA since its enactment. It is important that the actions and decisions of the courts be studied in detail, since to a large extent, the gaps left in the law by Congress have been filled in by these court decisions. To some degree, the law has been extended to encompass areas and activities, some of which were probably not intended to be covered by the act when the Congress passed it. Since the enforcement or action-forcing segments of the act were very broad, general, and, in some respects, vague, it is somewhat fortunate that in the very early days of the act, a multiplicity of court actions were brought against various Federal agencies and government officials by citizen environmental groups, and in some cases, by individual citizens. A close examination of these court decisions (particularly the language of the decisions) will give a good picture of the direction the act is taking and, more important, will provide knowledge of what the courts will require in the future as full compliance with the act.

The fact that NEPA, whatever the intention of Congress, was an overall reform of the procedural activities of many federal agencies, as well as a revision of much of the legal responsibilities of these agencies, and the fact that the Office of Management and the Budget declined to accept enforcement responsibility, make it almost obvious that implementation of the act must occur, in large measure, by interpretation from the judicial branch of the government. Whereas the Council on Environmental Quality was not delegated responsibility for enforcement, and the previously mentioned governmental establishments have declined to accept responsibility for the act's enforcement, the courts have openly accepted such responsibility. In some respects, federal judges seem to consider NEPA a valuable aid in their review of government policies. Their enthusiasm for NEPA in this regard may be seen in their interpolation, and, in many cases, extrapolation, of the spirit of NEPA (as they have interpreted the intent of Congress). This willingness, and sometimes enthusiasm, by the judicial branch of the government, to interpret NEPA may stem from the similarities between the ultimate goals of the act and the goals of judicial agency reviews.

Very broadly speaking, the early court actions and decisions after the passage of NEPA can be categorized in five general areas. These areas are: 1) requirement of formulation of systematic agency decision-making processes with respect to the preparation of an environmental impact assessment; 2) requirement of a full disclosure of the reasoning used in developing such procedures and in evaluating the risk associated with any final decision; 3) requirement of a full and complete evaluation of all possible alternative courses of action; 4) requirement of a more comprehensive assessment of what is the inherent interest of the public in the proposed action; and 5) requirement, heavily stressed in many recent decisions, of more public participation in decision making, with a further stipulation that the public be involved at a very early date. The courts have seized upon the opportunity of interpreting NEPA as a mechanism by which it may be possible to revamp and broaden judicial control over agency decision making, while simultaneously enhancing the human rights to life, health, and a healthful natural environment.

3-6.1 Judicial Interpretations of NEPA

Almost from the day that the National Environmental Policy Act was signed into law by the President, and, in some cases, even before the date when it became law, citizen interest in environmental conservation and public willingness to legally challenge almost any agency action have provided the courts with almost unlimited opportunities to play an exceedingly active role in the enforcement and interpretation of the act. While there are many court actions resulting from challenges of various segments of the act itself, by far the largest number of such actions, even from the very early days of the act, have related to Section 102(2)C, the requirement for an environmental impact statement. Many of the court decisions have been narrow in scope and have been confined to the action at hand with respect to a particular localized environmental challenge. However, there are several landmark cases that have greatly changed the act itself by setting precedents in the courts; this is especially important with regard to judicial definition of "full compliance" with the act. In this respect, it may not be an exaggeration to say that preparing an environmental impact statement today might be more closely related to satisfying the requirements of the various court decisions than to satisfying the requirements of the act as originally passed by Congress. Since this is true, it is important that these so-called landmark cases be described in some detail, in order to give the reader a better appreciation of the impact that a particular past court decision has had on the present and future requirements for total compliance with NEPA.

Since the enactment of NEPA in late 1969, the court decisions pertaining to the act, particularly those that have reached the United States Supreme Court, the United States Courts of Appeals, and the United States District Courts, can be grouped in five major divisions. These broad classifications are outlined below:

1. Since Congress, in passing the act, specifically expressed its desire to affect federal agency decision making at *all* levels and at *all* stages of the decision making, the courts have very actively concerned themselves with the lower limits of NEPA applicability (Ref. 3-19). In studying a project for long-term environmental consequences, the courts have determined that even projects of moderate size may have primary and secondary environmental impacts that may be quite significant. The courts have ruled that even localized, low-cost, small projects may warrant the integration of an environmental impact analysis into the early planning stages of the project.

2. Since one of the early mandates from the Council on Environ-

mental Quality was that all federal agencies are required to formulate specific guidelines to be used in their own agency decision making processes, the courts have held that each agency is responsible, at the very minimum, for the fulfillment of its own guidelines (Ref. 3-20). Several court decisions have required the preparation of an impact statement for an agency action, not so much on the grounds that the environmental impact would be significant, or because the action in question falls under the category of a "major" action, but simply because the published guidelines of the particular agency involved explicitly require the preparation of a statement for that particular type of action.

3. The courts have enumerated, and are constantly adding to the list, certain "typical" projects which must be the subjects of an environmental impact statement during their planning and development because of the obvious environmental consequences of such actions (Ref. 3-21). These types of projects have been identified collectively from various court decisions and agency actions, and from the many environmentally controversial projects which stimulated Congress to state an environmental policy for the nation.

4. As stated in the guidelines promulgated by the Council on Environmental Quality, all highly controversial projects must be described in an impact statement. The courts have interpreted the phrase "highly controversial" to refer to size, nature, or effect of federal actions, rather than to the existence of any organized opposition to a project (Ref. 3-22). However, the courts have also ruled that if published agency guidelines have defined "controversial" as meaning "meeting with strong public opposition," then that agency is bound by this interpretation in its own guidelines.

5. While the courts have generally reviewed impact assessments on a "case by case basis," in a few cases, the courts have recognized, in the search for precedents, the similarities among projects. It seems only logical that the frequency of such comparisons will increase in the future; this will have the effect of creating a trend away from the case-by-case judgement basis for each project, to what might be more

accurately referred to as a category-by-category judicial review (Ref. 3-23).

It can be seen that the courts have been playing a very active role in the interpretation and implementation of NEPA. Many of the court decisions handed down in the first few years after passage of the act have established what will probably be the foundations for the fundamental criteria to be satisfied in future environmental statements. These court decisions, as noted previously, have been addressed not only to items such as the general adherence by an agency to the requirements of the act, but also to the specific methods of preparation and detailed contents of impact statements as required by each agency's guidelines.

Many of the more important court decisions have been precedent-setting, with respect to the environmental impact statement itself. These particular cases are described in more detail in following chapters on the actual preparation of impact statements.

3-7. SUMMARY

It can thus be seen that what at first glance might appear to be an act conceived and enacted in haste, was, in fact, the culmination of more than a decade of legislative debate, compromise, and public pressures. The predecessors to Public Law 91-190, NEPA, were originally proposed to be only a statement of the nation's intention to preserve its environment from further degradation. Few, if any, with the possible exception of Senator Jackson, intended the bill to include the action forcing and enforcement provisions that were finally included at the time of the law's enactment. Most of the provisions of the law were opposed by the administrations in power during the decade, by Presidential advisors, and by many governmental agency staffs. Most of these opponents believed that the proposals were duplications of existing regulations and, as such, were unnecessary; yet the law made slow but steady progress toward passage.

Many of the major features of the present bill are the result of compromises made to move the bill along toward enactment. Among these features are the requirement for the environmental impact statement, the final form of the Council on Environmental Quality, and the lack of any agency being responsible or being charged with the enforcement of the bill's provisions. The conference committee also initiated the requirement for full inter-agency circulation of the EIS, as well as the requirement of a complete investigation of all possible alternates to the proposed action.

In spite of over ten years of debate and discussion, few in Congress realized the full impact of the law as it developed. It was just a matter of hours before the final passage of the bill in late December, 1969 that, as previously quoted, Representative Harsha stated to his House colleagues, "The impact of S.1075, if it becomes law, I am convinced would be so wide sweeping as to involve every branch of the government, every committee of Congress, every agency, and program of the nation." He very well might have added, in truth, that it would affect every citizen and business in the nation as well.

The law as finally enacted contains the three major components, 1) a statement of a national policy with respect to the environment; 2) the establishment of the three-man Council on Environmental Quality; and 3) the requirement of a detailed and comprehensive Environmental Impact Statement. However, what the act does not contain might possibly be more important in its long run environmental effect than what it does contain. It glaringly lacks specific details in regard to the disposition of an EIS, enforcement responsibility, and specifics of when a statement is required. Each of these omissions has led to many court interpretations of the law. Much of the present and future direction of NEPA has and is being determined by the courts. It is thus exceedingly important to discuss the major court decisions on the law and how they affect the required preparation of an impact statement. In the next chapter, Environmental Impact Statements are examined in detail along with these recent court decisions that bear upon their preparation.

REFERENCES

3-1. Boswell, Elizabeth M., *Federal Programs Related to the Environment*, U.S. Library of Congress, Legislative Reference Section, 1970.

3-2. U.S. Congress, House, Committee on Science and Astronautics, Subcommittee on Science, Research and Development, *Report, Managing the Environment,* 90th Congress, 2nd Session, 1968.

3-3. U.S. Department of Health, Education, and Welfare, *A Strategy for a Livable Environment, A Report to the Secretary of Health, Education and Welfare by the Task Force on Environmental Health and Related Problems,* (Washington: U.S. Government Printing Office, 1967).

3-4. U.S. Congress, Senate, Committee on Interior and Insular Affairs, *A Special Report: A National Policy for the Environment, Together with A Statement by Senator Henry M. Jackson,* 90th Congress, 2nd Session, 1968.

3-5. U.S. Congress, Senate, Committee on Interior and Insular Affairs, House, Committee on Science and Astronautics, *Congressional White Paper on a National Policy for the Environment,* 90th Congress, 2nd Session, 1968.

3-6. U.S. Congress, Senate, Committee on Interior and Insular Affairs, *Hearing, on S.1075, S.237, and S.1752,* 91st Congress, 1st Session, 1969.

3-7. U.S. Congress, House, Committee on Merchant Marine and Fisheries, Subcommittee on Fisheries and Wildlife Conservation, *Hearings on H.R. 6750 et al. Environmental Quality,* 91st Congress, 1st Session, 1969.

3-8. Finn, Terence T., *Conflict and Compromise: Congress Makes A Law, The Passage of the National Environmental Policy Act,* Georgetown University, Ph.D. Dissertation, p. 359.

3-9. Section 101(a), Amendment No. 25, S.1075, 91st Congress, 1st Session, 1969.

3-10. Section 102(3), Amendment No. 25, S.1075, 91st Congress, 1st Session, 1969.

3-11. U.S., *Congressional Record,* 91st Congress, 1st Session, 1969, CXV, Part 14, 19009.

3-12. U.S., *Congressional Record,* 91st Congress, 1st Session, 1969, CXV, Part 21, 29051.

3-13. U.S., *Congressional Record,* 91st Congress, 1st Session, 1969, CXV, Part 21, 29051–29052.

3-14. U.S., *Congressional Record,* 91st Congress, 1st Session, 1969, CXV, Part 30, 40416.

3-15. U.S., *Congressional Record,* 91st Congress, 1st Session, 1969, CXV, Part 30, 40928.

3.16. Finn, Terence T., *Conflict and Compromise: Congress Makes A Law, The Passage of the National Environmental Policy Act,* (Department of Government, Georgetown University) p. 425.

3-17. U.S. Executive Office of the President, Executive Order 11514, *Protection and Enhancement of Environmental Quality,* March 5, 1970.

3-18. U.S. Office of the Federal Register, *Federal Register,* Vol. 36, No. 79, pp. 7724–7729, *CEQ Guidelines on NEPA, Section 102 Environmental Statements,* April 23, 1973.

3-19. *Citizens for Reid State Park v. Laird,* 336F Supp. 783, 2ELR, 20122 (D. ME 1972).

3-20. *Scherr v. Volpe,* 336 F., Supp. 882, 2 ELR, 20068 (W.D. Wisc 1971).

3-21. *Indian Lookout Alliance v. Volpe,* 345 F. Supp. 1167, 3 ELS, 20051 (S.D. IA. 1972).

3-22. *Wilderness Society v. Hickel,* 325 F. Supp. 422, 1 ELR, 20042 (D.D.C. 1970).

3-23. *Virginians for Dulles v. Volpe,* 334, F. Supp. 573, 2 ELR, 20359 (E.D.VA. 1972).

4

The Environmental Impact Statement

4-1. INTRODUCTION

In the previous chapter, the requirement for the development of an environmental impact statement was described, as a part of the history of development of the National Environmental Policy Act. Certain conditions were detailed by Congress in the law, designating what types of Federal Agency actions would require environmental impact statements. In this chapter, it is our intention to answer many of the questions that could naturally arise, with respect to the contents of the actual environmental impact statement itself: questions such as, "Is an environmental impact statement needed for a particular action or project which an agency is about to undertake?" If, in fact, as EIS is needed, another question is raised: "What information must go into the statement and how much detail must be present in the study which develops the data base from which the statement is written?" If an EIS is written, or if it is decided that one is not necessary for the project, we ask: "What types of precedent-setting court actions affect the position taken?" If the decision is made to write an EIS, the following must be considered: "When in the course of the project planning shall it be written? When is it necessary to make the information available to the public? What are agency obligations after the draft statement is made public?" Finally, when the statement is complete: "Where is it sent and what is the obligation of the originator after the statement is filed?" These questions are answered in detail in the following sections.

It would appear that if impact statements are to serve their purpose, of complementing existing planning processes in the determination of future environmental effects, the statements must not reflect the off-hand opinions of the planner or developer, but rather, must result from a progressive and systematic evaluation of an assessment of the probable impacts of the project. The five points required by Section 102(2)C of NEPA, as referred to in the previous chapter, provide a general format for the contents of an environmental impact statement, but they do not provide the guidelines which are necessary for the procurement and analysis of the information required for the assessment process. Thus, it is necessary to establish a procedure for gathering and analyzing data in order to assess impact on the environment. Finally, it is essential that the data be presented in a form which is readable and understandable, not only to other technically trained people on the project, but also to the general public. This chapter is devoted to a discussion of some of the problems which will be encountered in developing a standard procedure or methodology for the impact assessment. Chapter 5 will discuss in detail the methodologies currently available and in use for gathering and assessing data.

4-2. IS AN IMPACT STATEMENT NEEDED?

The first question that must be answered when undertaking any action with the use of Federal Funds (which bring the action under the requirements of the National Environmental Policy Act) is, "Does the law require in this particular case that an EIS be written?" In answering this question, it is most important that one consider the intentions of Congress in enacting the law. As described in the previous chapter, the entire environmental impact assessment process has two major objectives. The first objective of the law is to make federal agencies more aware of the environmental impacts of their decisions and to bring to the attention of the decision makers the knowledge and additional information available from other agencies that may otherwise not have been used at the time the decision was made. The second objective of the law is to involve the public at an early stage of the planning of the project. This involvement should be such that, if needed, public pressure can be brought to bear upon the agency sufficiently early in the planning process to allow changes to be incorporated in the project which will reduce or eliminate the identified adverse environmental impacts. The EIS procedure itself can be an effective way for getting various concerned groups to work together. It forces the various federal agencies, local agencies, and the general public to participate more intimately in the planning and implementing of a project than was probable or than was required prior to the passage of NEPA. It will force inter-agency participation in major decision making, and will reinforce interaction between regional and project planning, particularly with respect to the environmental impacts.

4-2.1 Requirements of the Law

As stated in the previous chapter, Section 102(2)C of NEPA says that for every recommendation or proposal for legislation or other major federal action significantly affecting the quality of the human environment, a detailed statement by the responsible official must be made on: (1) the environmental impact of the proposed action; (2) any adverse effects which cannot be avoided, should the proposal be implemented; (3) alternatives to the proposed action; (4) the relationship between the short-term use of the environment and the maintenance and enhancement of long-term productivity; and (5) any irreversible and irretrievable commitments of resources, should the action be implemented (Ref. 4-1).

4-2.2   When is an Action Major or Federal?

The key phrase in Section 102(2)C, with respect to when an EIS is required, is: "major federal action significantly affecting the quality of the human environment." One of the first problems encountered in answering the question, "Is an EIS required?" is the ambiguity of the law itself. The federal agency official or statement developer first must determine what is a major federal action, and second, he must determine if it will significantly affect the quality of the human environment. But, what is a *major* action? This question has been a difficult one to answer. The guidelines of the Council on Environmental Quality, which will be discussed in detail later in the chapter, and which were issued to assist federal agencies in meeting the requirements of NEPA, have not been particularly helpful in this regard. The guidelines do not give specifics for determining when any action is major, or when an action is not a "major federal action" and is merely some other type of action. Second, the guidelines are not specific, with respect to defining the kinds of actions which will significantly affect the quality of the human environment. Thus, for the most part, the agency preparing the impact assessment has been left to its own devices or to guidance from the decisions of the courts to decide whether the action proposed will or will not require an environmental impact statement. It should be stated here that an environmental impact statement is not required every time a federal agency does anything. The determination of the circumstances which require an EIS ordinarily is made by the agency undertaking the action, but it should be remembered that this decision is subject to review by the courts. Ordinarily, a reviewing court will set aside the agency decision only if it finds the action by the agency to be "arbitrary and capricious, and to have abused discretion, or otherwise not in accordance with the law" (Ref. 4-2). The criterion for when an EIS is required was further amplified by the courts in the decision which has since become known as Hanly II (Ref. 4-3). In this case, the judges focused their attention on the phrase "significantly affects" as used in Section 102 of the law. In the decision the court stated:

. . . almost every major federal action, no matter how limited in scope, has some adverse effect on the human environment. It is equally clear that an action which is environmentally important to one neighbor may be of no consequence to another. Congress could have decided that every major federal action must therefore be the subject of a detailed impact statement prepared according to the procedure prescribed by Section 102(2)C. By adding the word 'significantly,' however, it demonstrated that before the agency in charge triggered that procedure it should conclude that a greater environmental impact would result than from just any major federal action.

Other cases in which the courts have rendered decisions relative to the requirement for an EIS are discussed in Section 4-2.4 of this chapter.

Another key factor in Section 102, relative to the requirement for preparation of an impact statement, is the use by Congress of the word "Federal." In the passage of NEPA, it was the intention of Congress to apply the law only to federal agencies. The compulsory preparation of an impact statement was not intended to apply to all the state, local, and private agency actions which may also cause environmental degradation. Thus, an important issue to be decided in answering the question of whether a statement is required or not is the decision of whether a particular action is sufficiently "federal" for NEPA to apply. This question may arise in the case where an action currently is, or at some earlier time has been, a state, local, or private action, but which is at the present, or which at sometime in the future, might develop into a federal task. In such a case, any environmental impacts which were caused might be construed as "federal," as intended by Congress in the passage of Section 102.

There are certain types of actions which, by their very nature, are federal in scope; thus, the agency carrying them out must prepare impact statements. Such actions as those by the Army Corps of Engineers, the Bureau of Reclamation, or by any of the branches of the Armed Services, would be so classified and, as such, would be covered by the requirements of NEPA. In such cases, ordinarily it is not necessary to discuss whether the federal presence is sufficient to justify a statement; it is assumed to be. At the same time, there are actions concerning the federal regulation of various activities (e.g., licensing or certification) which, if allowed, will obviously significantly affect the environment; this type of activity may qualify as major federal action. Cases such as the development of port authorities, marinas, and other similar waterfront projects requiring a permit from the Corps of Engineers would fall under this classification. The requirements of NEPA ordinarily have been applied to federal contracts, grants, and general loans to private parties as qualifying as federal actions by virtue of the use of federal funds. As a matter of fact, the very first case decided by the courts, with respect to the requirement of an environmental impact statement, was the case of *Texas Committee on Natural Resources v. United States* in which the court held that "there is little doubt that in the future the type of activity involved here would be covered by the statute" (Ref. 4-4). The court viewed as federal action not only the decision to grant a projected FHA loan (Farmers Home Administration Loan) for use in construction of a golf course and park (thus requiring an impact statement) but more specifically, the courts held that the entire project was a federal action. With respect to the phrase "major federal action," the precedent seems to be one in which the action is deemed major on the basis of the magnitude by which it significantly affects the environment, rather than upon the physical size or the dollar cost of the project itself.

4-2.3 Exemptions to Requirements of NEPA

Since the passage of NEPA, the courts have ruled in at least two cases that the requirement in the law for an impact statement on all agency actions does not apply. The first such ruling was one in which the courts held that the Environmental Protection Agency (EPA) need not prepare environmental impact statements for its environmentally protective activities (Ref. 4-5). This requirement has been challenged several times with respect to the implementation by the EPA of clean air standards and federal water pollution standards. However, perhaps because the EPA is charged by Congress with being the guardian of the national environment and responsible for setting limits on the discharge of various pollutants, the courts have ruled that this agency need not prepare an EIS on any of these actions.

The second exemption from the requirements of NEPA has been

made for an action involving national security (Refs. 4-6, 4-7). The courts traditionally have been reluctant to interfere in cases which involve national security or military installation projects. However, unlike the unanimous rulings with respect to EPA actions, court decisions have been quite divided with respect to the application of NEPA to the defense projects. In the case of *McQueary v. Laird*, the plaintiff asked the courts to require the Army to prepare a statement with respect to the effect of the storage of chemicals and bacterial warfare agents (Ref. 4-6). The court held that:

> . . . public disclosure relating to military defense facilities creates serious problems involving national security. We hold that NEPA does not create substantive rights in the plaintiff-appellants here to raise the environmental challenge in regards to the Rocky Mountain Arsenal in its proprietary military capacity, the federal government has traditionally exercised unfettered control with respect to internal management and operation of federal military establishment.

Thus the courts refused to require the writing of an impact statement.

Other cases, parallel in some respects, show similar reluctance on the part of the courts to interfere with military operations. For example, when the U.S. Navy planned to install an Ultra-Low Frequency Communication System over a thousand-square-mile grid in northern Wisconsin, the courts again refused to require that an impact statement be written.

In contrast to these decisions, the courts apparently had no difficulty in ruling that NEPA does apply to security-sensitive underground nuclear testing. In response to a challenge to one of the most controversial tests, that of Project Cannikin, which involved the testing of a five-megaton nuclear warhead in the Aleutian Islands off Alaska, the courts ruled that an impact statement indeed was required. However, in subsequent litigation, the challenge to the projected blast produced studies and data which cast greater doubt upon the environmental soundness of the test than did the Atomic Energy Commission (AEC) "official" impact statement. The courts were not at all clear in their ruling on this case and the U.S. Supreme Court ruling on this case left the issue even more in doubt. The majority of four in the Supreme Court merely denied the relief requested (that of stopping the AEC from the proposed test) without giving reasons or delivering an opinion. Perhaps the most positive statement that can be made, with respect to the exemption in national security cases, is that in agreeing to even hear the cases, the courts confirmed that they will police federal agencies such as the AEC, and that in cases of questions in regard to projects involving national security, federal agencies are not automatically exempt from the requirements of NEPA.

4-2.4 Other Court Decisions on Requirement for Statement

As pointed out in the previous articles, the law itself is somewhat vague and ambiguous in regard to when an EIS on a project is required. In view of this ambiguity, many of the initial court cases brought by citizen environmental groups and other interested parties had, as their primary plea, some form of statement regarding the requirement for an EIS on the proposed action. These early cases have set many of the precedents for determining if, in fact, Environmental Impact Statements are necessary for various federal actions, as interpreted in the requirements of NEPA. The landmark case in this regard is the *Calvert Cliffs'* decision (Ref. 4-8). This case, which involved a challenge to the AEC, has since been widely quoted as the yardstick against which the requirement for an EIS is to be measured. In this decision, Judge Skelly Wright issued a statement which has been repeated in several other court decisions. The decision as rendered stated:

> Section 102 duties are not inherently flexible. They must be complied with to the fullest extent, unless there is a clear conflict of statutory authority. Considerations of administrative difficulty, delay and economic costs will not suffice to strip the section of its fundamental importance . . . Compliance to the 'fullest' possible extent would seem to demand that environmental issues be considered at every important stage in the decision-making process concerning a particular action, at every stage where an overall balance of environmental and non-environmental factors is appropriate and where alterations might be made in the proposed action to minimize environmental costs.

In this same decision, the court defined the study of the relationship between the environmental, economic, and technical factors that must be considered, referring to this as a finely-tuned, systematic balanced analysis.

One other interesting and important court decision, relative to the requirement of an EIS, is that rendered in the case known as Hanly I or *Hanly v. Mitchell*. In this decision, the court analyzed the phrase in NEPA, "major federal action significantly affecting the human environment." In this case, the court ruled that the action must fulfill a three-part test to be considered qualifying under this particular statement in the law. The action must be major; it must be federal; and it must at the same time have a significant impact. An interesting aspect of this particular case is that in the decision, the court ruled that if the significant impact is predicted to be a beneficial one, then an impact statement need not be prepared (Ref. 4-9).

A second important trend in court decisions, regarding the requirement for an EIS, is one pertaining to the various agency guidelines required of all agencies by CEQ. These guidelines are discussed in detail later in this chapter. The courts have ruled that if an agency's guidelines cover the proposed action, or if the agency has begun or completed an impact statement, or otherwise conceded that a statement should be prepared, then in the court's view, these actions constitute an admission on the part of the agency that Section 102(2)C does, in fact, require that the agency in question write an EIS (for this particular project). In several cases, the decision has been rendered that, even though under other circumstances the project in question would not be considered a major federal action significantly affecting the environment, because the agency's guidelines have themselves suggested that a statement should be written, then a statement must be prepared. In this context, the courts have ruled that the federal agency's practices or guidelines have, in fact, lowered the threshold of "major" federal actions that would require the preparation of an impact statement. Some agencies have effectively lowered to zero this threshold of impact on which a statement is required. For example, the Federal Highway Works Administration has adopted the policy of preparing impact statements for practically every project which it approves or for which it provides funds. This thoroughness with which the FHWA has made highway projects a *per se* category for NEPA implementation, has then, in the court's eyes, been sufficient to require an EIS for every FHWA action. In the case of *Scherr v. Volpe* (Ref. 4-10), the FHWA attempted to reverse its policy of requiring impact statement preparation for small segments of highway. In this decision, the court refused to allow the FHWA to make an exception, citing, as justification, its own (FHWA) guidelines. In this respect, preparation of a statement has sometimes taken place even where the environmental impact is practically nonexistent. Another agency which has adopted a somewhat similar requirement is the Corps of Engineers. The Corps of Engineers, too, has been very thorough in implementing the requirements of Section 102(2)C, by requiring an EIS for almost every action, regardless of how small the project or how minimal the funds involved.

4-2.5 Timing of Statement—When Required

In the decision previously referred to in the *Calvert Cliffs'* case, the courts, in addition to ruling that an environmental statement was necessary, ruled that the requirements of NEPA were such that the impact statement must be prepared at the earliest possible time. In still another court decision, that of *Citizens for Clean Air v. Corps of Engineers* (Ref. 4-11), the court again stressed the point of early compliance with the requirements for an environmental impact study. In this decision, the court ruled that once a project has reached a coherent stage of development, the impact study must be undertaken. Federal judges have noted that one of the primary objectives of NEPA was to involve the public in decision making while alternate solutions are yet available. It seems clear from the various court decisions that have been rendered, that the intention of the courts and Congress is not to have the impact statement an after-the-fact document prepared when all of the planning and project preparation has been completed. On the contrary, it is their intention that the document be circulated to interested agencies and to the public at a date sufficiently early so that input into the planning for the project can be made and so that the project can be stopped, altered, or implemented, as the case may be, with full knowledge and input from all sources, in addition to the agency proposing the project. In particular, in the case of *Lathan v. Volpe* (Ref. 4-12), the courts have ruled that an after-the-fact EIS does not in any way satisfy the intentions of Congress or the requirements of the CEQ and the courts. In this case, the majority spokesman stated: "The defendants' suggestion that they will conduct additional research on its environmental effects after the highway is constructed, however, does not satisfy the requirements of NEPA. NEPA does not authorize the defendants to meet their responsibility by locking the barn door after the horses are stolen." Thus, the requirement that NEPA must be complied with before a major federal action is initiated precludes the preparation of the statement at the same time as, or after, the project goes forward.

Public participation at an early stage in project planning is one means of encouraging government and industry to be more responsive with respect to possible environmental improvement in their planning. Circulation of a draft EIS is the major means of public input to the assessment process. Even though often there is not a general distribution of draft statements (in most cases such statements must be specifically requested), once drafts are requested, the law requires that they be available to the public. This frequently leads the agency to not involve the public until after the planning has been completed. According to the CEQ guidelines discussed in detail later, the minimum comment period is to be 30 days with a 15 day extension if requested, after the draft statement is circulated, with a total delay of 90 days before any action can be undertaken. In any event, no action may be taken less than 30 days after the final statement is filed. However, effective public participation will not result if the agencies take a "wait for the public to come to us" approach. The agencies or their consultants need to establish a deliberate program to engage public interest at as early a date as possible in the assessment process. Flexibility in planning must be maintained for as long as possible before selecting that plan which will be finally recommended, so that the technical data from all sources, as well as the public input, concerning alternatives, may be properly analyzed. The courts, in their various decisions, have challenged the federal agencies to establish citizen participation in such a way that it will contribute significantly to the formulation of the proposed action and to the assessment of its impacts, as well as to the preparation of the EIS.

4-3. A DECISION THAT NO STATEMENT IS REQUIRED

As stated by Congress in the law itself, NEPA requires every federal agency to make the protection of the environment an intimate part of the planning for every project. The agency is not given the right to decide whether environmental values shall be considered, but rather is compelled to give the environmental impact of an action due consideration. The minimum requirement acceptable for compliance with the "action-forcing" provisions of the law is for the agency to issue a statement detailing the reasons why it believes that an EIS is unnecessary, when the agency has made the decision not to prepare such a statement. In this case, NEPA's requirement, as interpreted by the courts, is that the responsible official should prepare an environmental appraisal and issue a negative declaration. It is assumed that before issuing such a declaration, the agency would have prepared a reviewable record of the environmental impact to be expected.

Such a declaration by an agency, detailing the specifics of why a statement is not being prepared, will ensure that the agency has given adequate consideration to the consequences of its proposed action. In the case of *Sierra Club v. Hardin* (Ref. 4-13), which involved permits for timbering on federal lands, the courts asked specifically, with regard to the environmental effects, a very pointed question: "How can a decision be made whether to require a statement, until it is known just how significant the potential adverse impacts may be?" Thus, the judges saw very clearly that a declaration on the actual expected impact of a proposed action may, in fact, have to be prepared before a decision can be made adequately that no statement should be prepared at all. Going even one step further, in the case of *Natural Resources Defense Council v. Grant* (Ref. 4-14), the court stated, "An administrative agency may make a decision that a particular project is not major, or that it does not significantly affect the quality of the human environment, and that therefore, the agency is not required to file an impact statement." At the same time, in this same decision, the court cautioned that the agency may have to present its case and defend its decision not to write a statement for the proposed project.

One of the particular areas that has been neglected with regard to filing impact statements or even issuing a negative declaration is the matter of proposed legislation. Very few such legislative impact statements have been filed in NEPA's first six years of existence. However, the law specifically states that all agencies of the federal government shall include in every recommendation or proposal for legislation, as well as any other major federal action affecting the quality of the environment, a detailed statement. It has been estimated by one member of the House of Representatives that if the letter of the law was truly being followed, his committee alone would have to issue some 800 statements on proposed legislation during each session of Congress.

4-3.1 Minimum Requirements Without a Statement

Many cases were brought to the courts in the early days of NEPA, in which the primary claim was that an EIS actually was required even though the agency had issued a negative declaration. In these cases, there seem to be at least four criteria which the courts have used as yardsticks to determine whether a statement is required or not. The first of these is completion of a thorough investigation of the primary and secondary impacts of the action which, even though the project itself may be small, may warrant an environmental statement at the time of the project's planning. Another criterion the courts have referred to is the specific set of guidelines prepared by the agency at the request of CEQ. The courts have held that the agencies are, at a minimum, responsible for the fulfillment of their own guidelines. Therefore, as mentioned in Section 4-2.4, some court decisions have required the preparation of an impact statement solely on the grounds that an agency's policy would explicitly require such a statement. Third, the courts, as previously mentioned, have delineated certain typical projects which must have impact statements written, regardless of the agency's assessment of the impact. Finally, the CEQ guidelines have stated that all highly controversial projects (again regardless of the agency's statement of its evaluation of the impact), must be described in an EIS. Since NEPA passage in 1969, the courts have not hesitated to review the reasons given by an agency for issuing a negative declaration. They have, as a matter of course, subjected the declarations to searching reviews of the rationale by which the agency personnel have concluded that adverse affects are unlikely from a specific proposed action.

4-3.2   Court Decisions on Sufficiency of the Investigation

In the decision of *Hanly v. Kleindienst* (Ref. 4-3), the court analyzed an agency decision not to prepare an impact statement, and remarked that, ". . . in the context of an act designed to require federal agencies to affirmatively develop a reviewable environmental record, a perfunctory explanation of the agency's refusal to prepare such a statement would not suffice." The agency still had to show a record that would permit review by the courts, showing that it had properly considered environmental factors, although an impact statement might not have to be prepared.

The courts have ruled that for a very wide variety of cases, impact statements are not required. For example, statements were not required on such typical projects as: 1) the construction of a new U.S. postal service building on a 63-acre tract within a 350-acre industrial park; 2) the introduction of stretch jets into Washington National Airport in 1968; 3) the Interior Department's plan to fence 6800 acres of grazing land, acquired by condemnation, to protect a proposed reservoir; and 4) HUD's intention to insure a 3.8-million-dollar loan for the construction of a 272-unit apartment complex in a semi-rural neighborhood 15 miles south of Houston, Texas. Two items that the courts usually will review are how the agency gathered the information with respect to the impact of the proposed action and how the agency considered the information that they had so developed.

When in doubt as to the applicability of NEPA to a particular action which has been challenged by a citizen's environmental group, the courts appear to have taken the attitude that if doubt does exist or if public opposition is significant, an impact statement should be prepared. This general attitude by the courts is a major cause for several agencies, such as the Corps of Engineers and the Federal Highway Administration, adopting a policy that an EIS must be prepared for all but the most miniscule of projects. If this trend in court decisions continues, it seems fairly clear that more agencies will adopt such an attitude, and that, as time progresses, unless the law is amended by Congress, an EIS will have to be prepared for practically every action involving the expenditure of federal funds.

4-4. PREPARATION OF THE STATEMENT

According to Section 102(2)B of NEPA, "All agencies of the federal government shall identify and develop methods and procedures, in consultation with the Council on Environmental Quality, established by Title 2 of this Act, which insure that presently unquantified environmental amenities and values may be given appropriate consideration in decision making along with economic and technical considerations." Thus, in the same section (102) of the Act which requires that an EIS be prepared, Congress specifies that each agency is responsible for preparing guidelines detailing how this will be accomplished. Much of the apparent diversity of statements relating to projects of the same type may be attributed to this lack of clear definitive methods and criteria to quantify environmental impacts. Technical details, regarding how each agency will proceed in assessing impacts, are continually changing as the agencies gather more experience and, while gathering this experience, establish for themselves more competent data baselines, with respect to items in the environment, to which future impacts can be compared. The courts have held that Section 102(2)B does not, of itself, require the use of any particular technique for the environmental assessment, but is satisfied if the methodology used "effectively measures life's amenities in terms of the present state-of-the-art." The courts seemed to have been reasonable in allowing agencies or assessing consultants sufficient leeway for establishing procedures and details of the methods by which impacts will be measured and assessed.

4-4.1   Who Must Prepare the Statement?

Public Law 91-190 is quite clear in stating that the detailed statement required by Section 102(2)C must be prepared by a "responsible federal official." Much of the contention, with respect to this statement, has centered upon the question of the extent to which the federal official may allow the task of preparing the statement to be delegated to a consultant, a state agency, or to private parties. Another area of conflict has arisen with respect to the delegation of the responsibility in the case of two or more federal agencies sharing the authority for a project. In the early days of the Act, many of the agencies followed the practice of allowing the major responsibility for the preparation of the EIS to be passed along to the applicant in cases where the subject of federal action was the approval of a license, the granting of funds to a private party, or the funding of work to be carried out by a state or other federal agency. In this regard, the courts have ruled that the full responsibility and the burden for representing the public interest in the environment shall rest solely with the agency having primary responsibility for the action.

One agency's method of subrogating some of the responsibility for the preparation of the statement is that adopted by the Federal Highway Works Administration. The FHWA requires that the various states which receive federal trust funds for proposed highway construction prepare the required impact statements. The FHWA keeps this approach within the spirit of the law by having the "responsible federal official" accept the draft statements from the various highway departments, clear them for circulation, and provide for review and acceptance. The final statements must be signed and dated by this responsible official before they are officially approved by the agency.

A more recent trend has arisen in which agencies have allowed permit or license applicants to prepare the early drafts of the environmental statements, a policy which does not require the agency to bear all of the additional expenses to conduct the necessary studies. As long as the agency consults with the appropriate parties and then prepares the final detailed statement that describes the proposed action, the courts have seen fit to accept this approach as satisfying the requirements of the Act. It should be noted, however, that in this approach, the agency becomes vulnerable to some degree to a manipulation of the data used in determining which environmental impacts are considered to be significant.

With respect to those federal projects which may involve more than one agency, there are several means which seem to be acceptable with respect to fulfilling the requirements for responsibility in a federal official. Among these several alternate means for complying with NEPA is one in which each agency would prepare a statement on its specific part of the action, with all of the separate statements being issued as the summary statement for the project. Another approach is for the agencies involved to jointly prepare a statement with the combined statement being issued in the name of all the agencies. A third approach is for a single agency to take the responsibility of preparing the complete assessment and impact statement, in spite of the fact that other agencies are involved in the proposed action. Of these various methods, the preferred methods, under the current guidelines, seem to be the preparation of individual statements by each agency or a joint preparation participated in by all of the agencies. Some difficulty has been experienced with both these approaches, in overlapping statements or in having some areas of environmental impact not assessed or evaluated by any of the agencies involved in the project in question.

4-4.2 Sequence and Timing

The general sequence of events in preparing an EIS within a federal agency, as might be expected, will vary somewhat from agency to agency. In general, however, the first step is for the applicant agency to prepare an environmental assessment of the proposed action, after collecting all the baseline environmental data required for this assessment. The agency then performs an environmental review of its actions at a very early stage in the overall project development, in order to decide whether or not an EIS is required for the particular project under consideration. If it is determined that a statement is required, a more detailed review of the predicted impacts, compared to the previously gathered baseline data, is made. In this case, the law is quite specific with regard to the items which must be addressed, as pointed out in the discussion of NEPA in the previous chapter. Such items as the positive and negative effects of the proposed action on the environment, alternates to the action, probable adverse environmental effects which cannot be avoided, and commitments of resources are a few of the factors which the law requires to be evaluated.

Recent CEQ guidelines have required the draft statement to detail the purpose of the action, its relationship to other existing plans and policies, and any anticipated secondary environmental effects. Impact statements now are also required to contain alternate designs or details which would significantly conserve energy.

The agency is required to furnish impact statement materials to the public without charge, if possible; if a fee is necessary, the fee cannot be more than the actual cost of reproduction.

The minimum period which an agency must allow for review and comment on its draft impact statement is currently 45 days. However, the length of the comment period must be established on the basis of the complexity and size of the impact statement, along with the extent of public interest in the proposed action. If a longer time is required, the agency must allow additional time for review when a reasonable request for such an extension is received.

When the comment period is over, the agency must address the questions and unresolved issues raised during the comment period by other agencies and the general public. The draft statement then is amended as necessary, with the comments attached (along with replies and necessary supplementary data), and is forwarded as the Final Impact Statement.

4-4.3 Draft and Final Statements

As described in the previous section, the responsible federal offical is required to prepare a draft impact assessment or statement to circulate to appropriate agencies, the EPA, and to citizen groups who request copies. Later, this draft, all comments, and, if necessary, more detailed baseline data, are combined in a final impact statement. The main differences between the draft and final impact statements are in purpose and contributors. Draft statements are prepared by the lead agency or by consultants delegated this responsibility by the agency, to present its ideas of impacts resulting from the proposed action or alternates to the action. The final impact statements include the lead agency's discussion, along with comments from federal, state, and local agencies, public organizations, and private citizens. The purpose of the final statement is to present a full disclosure of the impacts of the proposed project and all alternatives *as conceived by all concerned parties*. The final statement must be comprehensive and detailed, with explicit statements of all the predicted primary and secondary impacts on the social components of the human environment, as well as on the physical and biological systems.

The following elements appear to be critical to a draft EIS: 1) a complete inventory and objective analysis of all environmental impacts associated with the proposed project, including both beneficial and adverse impacts; 2) a detailed statement of the adverse impacts of the project and what, if any, remedies are available to alleviate them; 3) a statement giving the details of project compliance with existing environmental law; and 4) a listing of the possible alternatives to the project, including a so-called null (or do nothing) alternative. Each alternative should be analyzed from an engineering and economic point of view, as well as with respect to environmental costs.

The most important elements to be included in the final EIS are: 1) a chronological listing of all communications with other federal agencies and the public, both before and after the issuance of the draft statement; 2) an examination or transcript of the public hearing, if one is held, with a discussion of the responses to the complaints received; 3) a reply to all comments and questions received during the review period; and 4) an unbiased presentation of all unresolved problems associated with the draft statement.

The final environmental statement must be a decision-forcing document that has been fully scrutinized and criticized by federal, state, and other agencies with expertise relative to the project, as well as by the public sector. When one considers what should be included in an EIS, there are many sources that can be investigated for guidelines. Among the considerations are the requirements of NEPA itself, the requirements as delineated by CEQ, and those guidelines of the agencies required to prepare impact statements.

However, perhaps one of the most concise statements summarizing the philosophy of what should be in the statement, as required by NEPA, was given by a judge in his decision on the Gillham Dam case (Ref. 4-2). In this decision, Judge Eisele concluded, "At the very least NEPA is an environmental full disclosure law. The detailed statement required by Section 102(2)C should, at a minimum, contain such information as will alert the President, the Council on Environmental Quality, the public and, indeed, the Congress, to all known possible environmental consequences of proposed agency action."

4-5. CONTENTS OF AN EIS

A usual complaint made by groups bringing challenges to impact statements released for public comment concerns the lack of specifics in the statement. Two decisions were directed toward these complaints in the early days of NEPA. The first of these was the now famous Calvert Cliffs' decision which provided the opinion that the environmental analysis must be made to the "fullest" extent possible before the statement could be considered to be in full compliance with NEPA. Second, a phrase occurs in the CEQ guidelines, discussed later in this chapter in more detail, in which the Council member states, "The amount of detail provided in an environmental statement should be commensurate with the extent and expected impact of the action, and with the amount of information required at the particular level of decision-making (planning, feasibility, design, etc.)." These guidelines also state that the lead agency should make every effort to convey the information in a form easily understood, both by members of the public and by public decision makers, giving attention to the substance of the information, rather than to the particular form, or length, or detail of the statement.

4-5.1 Requirements of NEPA and CEQ

The requirements of NEPA, as previously stated, are summarized in the five points specifically detailed in the law. To these five points, as itemized in Section 102(2)C of the Act, are sometimes added several points of coverage from the CEQ guidelines. These points require certain items or information to be provided. The first point requires a description of the quality of the environment, including a description of aquatic and terrestrial ecosystems. This point requires a detailed description of the proposed action. The description must include specifics with regard to the area of the project, the resources involved, the physical changes proposed, and the ecological systems to be altered. This point also requires a description of the environmental interrelation, if any, of the particular project with the total affected area, however extensive that area may be.

The second point requires a description of any probable impact on the environment, including, specifically, the impact on ecological systems (e.g., species, habitats, communities). The CEQ guidelines, in elaborating upon this point, state that the significant actions to be addressed here are to be interpreted to include those which may have both beneficial and detrimental effects, even if upon balancing these effects, the agency believes that the *overall* effect will be beneficial. Thus the agency proposing the action must consider and report all alterations to existing conditions, whether they are deemed beneficial or detrimental.

The third point requires the responsible agency to study, develop, and describe appropriate alternatives to the recommended courses of action in any proposal which involves unresolved conflicts concerning alternate uses of available resources. Sufficient analysis of such alternatives and their cost and impact on the environment must accompany the proposed action through the agency review process, so that options which may have a less detrimental effect are not foreclosed. This requires not only a discussion of complete alternatives which would accomplish the objective with less impact, but also requires a discussion of partial alternatives and those that would include the elimination of certain ''high environmental impact'' aspects of the proposed action. Most proposed actions involve a number of potential areas or segments where an imaginative approach may lessen an adverse impact while satisfying the basic requirements of the project at the same time. An environmental statement should describe these alternatives in such a manner that a reviewer of the statement can independently judge whether the particular aspect should be eliminated or not.

Point four requires an assessment of the relationship between the local short-term uses of man's environment and the maintenance and enhancement of the long-term environmental productivity. The agency or consultant is thus required to assess the proposed action for cumulative and long-term effects on the environment. The project or action must be evaluated in terms of the use of both renewable and non-renewable resources. In effect, the statement must delineate who is paying the "environmental cost;" the people who presently gain the benefits from the project or future generations who may be left with only the cost.

Point five requires a description of any irreversible and irretrievable commitments of resources. For example, construction of a storage reservoir, construction of a highway or pipeline, and dredging of marshlands all would, for all practical purposes, involve an irretrievable commitment of land resources. Irreversible damage might also result from accidents; for example, from an oil spill. The risk and possibility of such occurrences as these should be fully discussed.

Finally, other points, while not occurring in the law as enacted by Congress, have been added by the guidelines of CEQ. For example, discussion, where appropriate, of the problems and objections raised by local entities during the review process, must be included. The purpose of this requirement is two-fold. It encourages the agency proposing the action to contact and communicate with the various interested groups during early planning and throughout the project development. It also provides reviewers with reference or contact with groups who may have personal knowledge of the impact of the proposal. Other points required by the CEQ guidelines are described below.

In addition to the previously discussed requirements of the Act itself, relative to the contents of an environmental impact statement, the law further directed the Council on Environmental Quality to issue guidelines for agency action as required by Public Law 91-190. On August 1, 1973, CEQ issued guidelines for the preparation of environmental impact statements. These guidelines appear in the Code of Federal Regulations, Title 40, Chapter V, as Part

1500. In these regulations, Section 1500.8, entitled "Content of Environmental Statement," gives a detailed description of the points required by CEQ to be covered in a statement. Specifically, the guidelines cover eight individual items which must be addressed fully in any impact statement. However, the guidelines further state that in developing these eight points, agencies should make every effort to convey the information in a form easily understood by the public. The guidelines also state, as a matter of clarification, that each of the points need not always occupy a distinct section of the statement, if the point is otherwise adequately covered in discussing the impact of the proposed action and its alternates. These items, of course, should normally be the focus of the statement itself.

In summary, the CEQ guidelines require: 1) A description of the proposed action, a statement of the purposes of the action, and a description of the environment affected; this section should include ecological information, summary technical data, maps, and diagrams where relevant, all of which must be adequate to permit an assessment of the potential environmental impact of the project; 2) An explanation of the relationship of the proposed action to the current land use plans, policies, and controls for the affected area or region; 3) A statement of the probable impact of the proposed action on the environment; this requires the agencies to assess both the negative and positive effects of the action as it affects both the national and international environment. This requirement is similar to NEPA requirements, with the specific addition that secondary or indirect, as well as primary or direct, consequences for the impacted environment should be included in the analysis; 4) Alternatives to the proposed action, including, where relevant, those not within the existing authority of the responsible agency. The agency is required to make a rigorous exploration and an objective evaluation of the environmental impact of all reasonable alternative actions, particularly those that might enhance environmental quality or avoid some or all of the adverse environmental effects; 5) Discussion of any probable adverse environmental effects which cannot be avoided (such as water or air pollution, undesirable land use patterns, damage to life systems, urban congestion, or other consequences adverse to the environmental goals set out in Section 101 B of the Act); 6) A description of the relationship between local

Fig. 4-1 (a) Power plant and electrical generation station.

Fig. 4-1 (b) High voltage transmission lines from power plant to urban areas. In environmental assessments of energy production and transmission activities, for example, how far should the investigation be carried? Along transmission lines? To the power plant? To the coal mine?

short-term uses of man's environment and the maintenance or enhancement of its long-term productivity; 7) An evaluation of any irreversible and irretrievable commitments of resources that would be involved in the proposed action should it be implemented; and 8) An indication of what other interests and considerations of federal policy are thought to offset the adverse environmental effects of the proposed action. The statement should also indicate the extent to which these stated countervailing benefits could be realized by following reasonable alternatives to the proposed action, that would avoid some or all of the adverse environmental effects. In this connection, agencies that prepare cost/benefit analyses of proposed actions should attach such analyses or summaries thereof to the EIS, and should clearly indicate the extent to which environmental costs have not been reflected in such analyses.

The purpose of these guidelines, as stated, is to assist the various federal agencies in preparing their own individual guidelines, so that NEPA may be implemented through the cooperation of all federal agencies. In essence, they are a detailed expansion of NEPA provisions. The guidelines attempt to require early assistance and closely integrated cooperation between each agency and CEQ, so that the impact statement preparation can become a major factor in the decision making process, rather than an after-fact justification for a project, based upon traditional decision making premises. It should be pointed out in this regard that the three main areas of environmental impact identified by the CEQ guidelines are pollution, energy supply and natural resources development, and land use and management. Each of these three main categories is then further subdivided into impact types with a listing of which agencies are responsible for controlling the various types of impact so identified. The reader is referred to the complete CEQ guidelines in the Appendix, for more detail.

4-5.2   Other Agency Guidelines and Examples

As a result of the passage of NEPA, and pursuant to Section (2) of Executive Order 11514, all federal agencies were directed to proceed with any measures necessary to implement Section 102(2)C of NEPA. As a result of these actions, CEQ, in publishing its own guidelines, entitled "Statements on Proposed Federal Actions Affecting the Environment," published in the Federal Register of April 23, 1971, directed each federal agency not only to comply with the CEQ guidelines, but also to formulate their own. Section three of the CEQ guidelines states, "Each agency will establish . . . not later than July 1, 1971 with respect to requirements imposed by these guidelines, which apply to draft environmental statements circulated after June 30, 1971, its own formal procedures for (1) identifying those agency actions requiring environmental statements, (2) obtaining information required in their preparation, (3) designating the officials to be responsible for the statements, and (4) consulting with and taking account of the comments of appropriate federal, state and local agencies." In other words, each federal agency was directed by CEQ to submit detailed guidelines showing not only how that particular agency would prepare statements on their own actions, but also how they would review and comment upon draft environmental statements received from other agencies. As might be expected, because of the enormous diversity in the types of federal actions and projects controlled by the various agencies, the guidelines differ significantly from agency to agency. In addition, since the courts are constantly handing down decisions relative to NEPA, as would be expected, the guidelines frequently are amended or changed by the various agencies. In the preparation of an environmental assessment, one of the first steps is to check the latest editions of the guidelines of the various agencies that will review and comment upon the draft statement as prepared and circulated. As an example of the types of guidelines which the various agencies have circulated, the following guidelines are summarized from those submitted by Region X of the Environmental Protection Agency, dated April 1973 (Ref. 4-15). The particular guidelines selected for this example are those detailed for preparing or reviewing a draft impact statement for a planning unit, or an action which will develop the long-range use of a specific land tract.

REGION X EPA GUIDELINES—PLANNING UNIT

Planning Units

A planning unit describes long-range multiple use objectives and policies for a specific land tract, including the allocation and values of resources. The plan provides guidance to a manager of an area based on existing inventories and knowledge of how a land tract can be managed, utilized, and protected. It addresses such resources as recreation, timber, watersheds, mining, and wildlife. The compatibility of these resources to each other and to existing conditions defines the suitability of activities that will be allowed. A planning unit may be considered as an initial concept to be developed into descriptions of proposed activities.

Once detailed planning for an area has been devised, an environmental impact statement can be prepared. Specific details of the proposed land use plan and its probable impacts are essential in the development of an environmental impact statement.

Planning Unit—General Guidelines for Impact Assessment
I.   *Description*
    A. Describe in qualitative and quantitative terms all biological and water resources. This discussion should include how the biotic communities have adopted to the physical environment, and should also include the hydrologic cycle of adjacent water bodies.
    B. Describe the soil characteristics and geology in the project area.
    C. Describe all natural resources in the project area, including wilderness areas. The statement should recognize that these wilderness areas are a diminishing resource.
    D. Describe existing air quality and any applicable standards or regulations.
    E. Include graphic and pictorial information.
    F. Describe meteorological conditions in the area.
    G. Describe past, present, and proposed land use.
    H. Describe accessibility to the planning area and include any short and long-range transportation plans.
II. *Environmental Impacts*
    A. Discuss any impacts which may occur to water quality or air quality, as well as impacts from noise, solid waste disposal, and pesticide use.
    B. Discuss the impacts that the project will have on the physical environment such as soils, geologic formations, hydrology, drainage patterns, etc.
    C. Discuss methodology to be used to minimize any adverse environmental impacts.

III. *Alternatives*

A. Discuss the full range of management alternatives, including the "null" alternative, considered in the course of planning the action.

B. Discuss in depth why the proposed alternative was chosen.

C. Discuss alternatives in sufficient detail so reviewers may realize secondary or long-term environmental impacts.

IV. *Short-Term Uses vs. Long-Term Productivity*

A. Discuss environmental cost as it relates to short-term uses and long-term productivity.

B. Discuss how actions taken now will limit the number of choices left for future generations.

V. *Irreversible and Irretrievable Commitment of Resources*

Discuss resources to be utilized and what the replacement potential of these resources is.

REGION X EPA GUIDELINES—RECREATION AREA

Recreation Areas

Recreation is one of the components to be considered in multiple-use forest management. The term "recreation area" refers to a broad range of uses, from preservation and enjoyment of the natural environment (as provided by the National Wilderness and Wild and Scenic Rivers Systems) to highly developed recreation areas, such as ski resorts. It is the latter type of recreation area which places the greatest stress on the environment and which requires the inclusion of careful planning and environmental safeguards to preserve the delicate balance between man and nature. The following guidelines refer to the environmental assessment of those recreational areas such as campsites and winter/summer resorts.

I. *Description*

A. Illustrate the proposed facility and the topography of the area by including graphs or pictures.

B. Describe the geology and soil characteristics of the area.

C. Describe the biotic community. List existing types of vegetation and animals, including their present ecological relationship. Identify any rare or endangered species in the area.

D. Describe the quantity, quality, and characteristic uses of water bodies in the area.

E. Describe the location and placement of any facilities associated with the project, i.e. base facilities, life terminals, and parking lots.

F. Describe the expected uses of the new facility. Discuss any seasonal uses. Will recreation vehicles be allowed?

G. Describe potable water supply and predicted demand.

H. Describe the methods included in the project to provide waste treatment. Sufficient information must be provided to determine whether the planned treatment facility will provide the waste treatment necessary to prevent stream degradation. Indicate the design capacity of any planned treatment plant. In view of the seasonal, transient nature of the users of recreational facility, how will adequate waste disposal be maintained under such varying load conditions? Will emergency storage be necessary to avoid spillage into adjacent water bodies if the treatment facility fails?

I. Discuss the ultimate disposal of wastes generated by users of the facility and indicate the types and volumes of such wastes. Consideration of disposal methods should include all practicable methods to dispose of liquid and solid wastes and should be considered in sufficient detail to allow the reviewer to decide whether adequate environmental protection features have been included.

II. *Environmental Impacts*

A. Discuss the impacts which will occur to the ecosystems in the project area.

B. Address all impacts on water and air quality due to construction and maintenance of the proposed project.

C. Discuss the impacts of noise generated by the project. Include present and predicted noise levels due to the proposed activity.

D. Discuss impact on soils and geology in the area. Details should be included on the possible effects that soil erosion and turbidity will have on existing water bodies and the affected ecosystem.

E. Discuss the secondary impacts accompanying the proposed project. Recreation areas create many new demands in and around an area; discuss any predicted impacts as a result of these demands.

F. Describe all controls to be incorporated into the project to prevent or reduce any environmental impacts. Specific details should be included on the method to prevent soil erosion, excessive noise levels, and air and water degradation.

III. *Alternatives*

A. Discuss the no action or "null" alternative.

B. Discuss alternative locations and impacts for the proposed project.

C. Discuss alternatives related to the size and magnitude of the proposed action.

D. Discuss alternate locations of facilities which might lessen the environmental impacts.

IV. *Short-Term Uses vs. Long-Term Productivity*

A. Address the short-term benefits and relate them to environmental losses.

B. Discuss the environmental trade-offs involved in converting an area to a single purpose resource.

C. Discuss any gains from long-term productivity of the area.

V. *Irreversible and Irretrievable Commitment of Resources*

Discuss the quantity, quality, location, and accessibility of those renewable and non-renewable resources to be utilized by the proposed action. For example, converting a multiple purpose area to a single purpose use may be an irretrievable commitment of the land.

4-5.3 Adequacy of the Statement

As might be expected with an issue as emotional, from the public's point of view, as the preparation of an EIS on a federal agency action that will affect the lives and the environment of many sectors of the public, many court actions have been directed toward judging the adequacy of the EIS circulated for public comment. The courts have ruled on this specific question, handing down decisions in several cases. The question, from the judicial point of view, has been somewhat delicate in that the courts have been reluctant to actually take a stand and render a decision on the basis of the adequacy of the technical and scientific data presented in the statement itself. The judiciary has acted to balance the need for complete environmental disclosure against the reasonableness with which resources may be devoted to the environmental investigation. For example, in the litigation with regard to the Tennessee-Tombigbee Navigation Project, the courts stated, in the case of *Environmental Defense Fund v. Corps of Engineers* (Ref. 4-16): "In reviewing the sufficiency of an agency compliance with Section 102, we do not fathom the phrase 'to the fullest extent possible' to be an absolute term requiring perfection. If perfection were the standard, compliance would necessitate the accumulation of the sum total of scientific knowledge of the environmental elements affected by a proposal. It is unreasonable to impute to the Congress such an edict. The phrase 'to the fullest extent possible' clearly imposes a standard environmental management, requiring nothing less comprehensive than an objective treatment by the responsible agency. Thus an agency's consideration of environmental matters that is merely partial or performed in a superficial manner does not satisfy requisite standard."

In determining whether statements are reasonably adequate, the courts have indicated that 1) they must be understandable and non-conclusive; 2) they should refer to the full range of knowledge currently available on the topic; and 3) they must discuss all impacts which are typical of certain types of action. In the decision referred to above, the courts further stated that the impact statement should, within reason, be as complete as possible. Yet there would be no opposition to an agency bringing new or additional information, opinions, or arguments to the attention of the decision makers even after the final EIS has been forwarded to CEQ. It is doubtful that any agency, however objective and well staffed, will ever come up with a perfect environmental impact statement in connection with any major project. It seems almost obvious that, regardless of the time and effort put into preparation, detailed anaylses by teams of experts in specific areas (such as will be reviewing the draft statement in various agencies) are almost certain to reveal inadequacies or deficiencies. However, it would seem that one of the purposes for the circulation of the draft statement to all concerned agencies is to reveal these deficiencies at as early a date as possible. Thus, the final statement, when forwarded to CEQ, must include all agency and public comments along with the proposed answers or elaborations to such comments.

Another area related to this problem of adequacy is the method of handling uncertain or unknown impacts. Since the impact statement is prepared before the action is carried out in most cases, the prediction of the impact is, in large measure, based upon informed guesswork. For some types of action, the range of uncertainty may be quite serious, as in the case of actions or projects which are the first of their kind. In other situations, the effort necessary to measure the precise impacts may be excessively costly. The courts have differed in their approaches to this problem. Some have found adequate compliance with NEPA when the EIS simply points out that a gap exists in the current knowledge on the particular item. Other courts have made the completion of an agency research program a prerequisite before beginning any action on the project. It has been suggested that the EIS should be considered sufficient if the decision makers are alerted to the problems.

4-5.4 Examples of Tables of Contents of Completed Statements

As examples of listings of the types of impact areas discussed in environmental statements, the following two tables of contents, from actual EIS filed with CEQ, are included as perhaps typical of an EIS for a small or medium size project. The first is for a river navigation project, while the second is for an industrial port on a major navigable river.

DRAFT ENVIRONMENTAL IMPACT STATEMENT
CLEAR RIVER NAVIGATION PROJECT

OPERATION AND MAINTENANCE

TABLE OF CONTENTS

Section   Title

1   PROJECT DESCRIPTION
       LOCATION AND PURPOSE OF THE PROJECT
       PROJECT ELEMENTS
         Authorization
         Costs
         Structural and Operational Components

TABLE OF CONTENTS (Continued)

TABLE OF CONTENTS (Continued)

TABLE OF CONTENTS (Continued)

DRAFT ENVIRONMENTAL IMPACT STATEMENT
MARITIME CENTRE—INDUSTRIAL PARK

TABLE OF CONTENTS

TABLE OF CONTENTS (Continued)

TABLE OF CONTENTS (Continued)

4-6. ALTERNATES TO THE PROPOSED ACTION

The wording of the law is quite specific, in at least two places, in saying that any agency proposing an action or project must consider alternate methods or procedures for the proposed action. In Section 102(2)C, the text states that a detailed statement must be made by the responsible official on alternatives to the proposed action. In the same section, paragraph D, the requirement is stated that all agencies must "study, develop, and describe appropriate alternatives to recommended courses of action in any proposal which involves unresolved conflicts concerning alternate uses of available resources." The positiveness with which Congress has stated that alternate solutions must be considered, in the case of any adverse environmental impacts, has lead to a number of early court decisions. In these actions, the courts have considered rather carefully whether or not the agencies have in good faith considered alternate methods of completing the proposed action. One of the early decisions in this respect was the case involving the Gillham Dam, mentioned previously, in which the EIS was prepared by the Corps of Engineers. In this case, the court recognized that there was an alternate of no action at all, or of another action that fully accomplished the original goal but without many of the objectionable features of the project, and that between these extremes probably other alternatives existed. The court, in its decision, characterized the Corps' failure to describe fully the consequences of leaving the river alone, that is, taking no action whatsoever, as a "glaring deficiency" in the statement. In the decision in the case of *Natural Resources Defense Council v. Morton* (Ref. 4-17), the court went one step further when it stated ". . . nor is it appropriate, as government counsel argues, to disregard alternatives merely because they don't offer a complete solution to the problem. If an alternative would result in supplying only part of the energy that the lease sale would yield, then its use might possibly reduce the scope of the lease sale program and thus alleviate a significant portion of the environmental harm." In this same decision, the court ruled that the agency must discuss all reasonable alternatives to the proposed action, whether or not such alternatives come within the direct control of that particular agency. In addition, the court stated that the alternates must be discussed in sufficient detail to allow a reviewer to make a reasonable choice. The courts thus ruled that the impact statement should not only mention the alternatives, but should attempt to compare the impacts of each alternative to the main proposal.

In the *Calvert Cliffs'* decision, the court stated, "The detailed statement requirement seeks to insure that each agency decision-maker has before hand, and takes into proper account, all possible approaches to a particular project (including total abandonment of the project) which would alter the environmental impact and the cost-benefit balance. Only in that fashion is it likely that the most intelligent, optimally beneficial decision will ultimately be made."

As mentioned, Section 1500.8(a)(4) of the CEQ guidelines is quite explicit in its requirement of the treatment of alternatives. This guideline states, "A rigorous exploration and an objective evaluation of the environmental impacts of all reasonable alternative actions, particularly those that might enhance environmental quality or avoid some or all of the adverse environmental effects is essential. Sufficient analysis of such alternatives, and their environmental benefits, costs, and risks, should accompany the proposed action through the agency review process in order not to foreclose prematurely options which might enhance environmental quality or have less detrimental effects."

Many agencies and consultants are finding this requirement of NEPA difficult to accept since the traditional engineering report usually concentrates on explaining only what the agency considers the "best plan" to accomplish the desired end. The philosophy of an engineering report historically has been not to discuss possible alternate ways of achieving the project, but to present what the agency or the consultant considers the most feasible way of accomplishing the objectives. The fact that alternative choices for achieving the desired end are not discussed does not mean that choices are not made. In fact, in coming to a recommendation on what the agency considers the best plan it is imperative that alternates be considered. However, historically, the alternative choices have been hidden; ordinarily, they are not explained in the engineering report. It is probably for this reason that many of the early environmental impact statements were rejected by the courts as inadequate; the alternatives that were, in all probability, considered in making the decision to accomplish the project in a particular way were not included.

One of the most interesting agency guidelines in this respect is that promulgated by the Environmental Protection Agency which requires the agency to ". . . develop, describe, and objectively weight alternatives in any proposed action which involves significant tradeoffs among the uses of available environmental resources. The analysis shall be structured in a manner which allows comparisons of: 1) environmental and financial cost differences among equally effective alternatives; or 2) differences in effectiveness among equally costly alternatives. Where practicable, benefits and costs should be quantified or else described qualitatively in a way which will aid in a more objective judgment of their value" (Ref. 4-18).

As the courts have broadened the requirement for considering possible alternatives to an action, a geometric progression in the number of possible alternatives threatens to overwhelm the agency or the consultant. This is particularly true since, as reported above, the courts would seem to require not only alternates to the entire project, but alternates to various pieces that may accomplish much less than the original proposal. Some way of reducing this discussion of alternates to a reasonable number would therefore seem to be desirable. On the other hand, it does not seem wise to allow the agency to have the discretion to pick the alternatives to be discussed. If such discretion were allowed, it is obvious that almost any proposal could be made to look better than the alternatives then presented and discussed. Cognizant of this dilemma, the courts ruled in the case of *Natural Resources Defense Council v. Morton* (Ref. 4-17) that the "rule of reason" be used to limit consideration of alternatives to those available in the same time span as the original proposal, and further, that only reasonably approximate alternatives are to be discussed, excluding those which involve the repeal of "basic legislation."

4-7. FULL DISCLOSURE AND PUBLIC COMMENT

One of the still unresolved issues, with respect to the requirements of NEPA, is the decision with regard to whether or not the impact statement should be able to serve as the full record of the final agency decision, or whether it is only a part of that record. If the statement is to serve as the full record available to the public and commenting agencies, then it must include not only the conclusions, but the data and reasoning from which the conclusions were drawn to support the agency's action. One of the purposes to be considered in preparation of the impact statement is that it should serve as a full disclosure of all substantial environmental effects of a proposed action, and how these will affect the quality of the human environment. As previously stated, it has been agreed by the courts that, at the very least, NEPA is an environmental full-disclosure law. Further, the law itself states that as a minimum, it should contain information that will alert the President, CEQ, Congress, and the public to all known possible environmental consequences of the proposed action. In addition to being spelled out in the law itself, the requirement for early and complete public participation in the impact statement process has been cited many times by the courts. The public must be allowed to comment when the agency is deciding whether or not a particular action will reach the threshold level for environmental harm, and should be involved actually in the decision as to whether or not an impact statement will in fact be prepared. In this respect, it might be stated that NEPA is, in fact, a unique statute. The Act creates an important new public right to be informed of the possible environmental consequences of agency actions, to have alternatives considered, and to have the interest of future generations taken into account. NEPA grants to the public the right to participate in decision making. The public must receive adequate notice of the proposed action and, as a matter of fact, the statement itself must show how the public was allowed to participate in the actual decision. One of the interesting sidelights of the decisions that the courts have made, relative to NEPA, is the stand taken with respect to industrial plaintiffs. The courts have not, as yet, ruled that industry must be allowed to participate in the planning and decision making in the same way that the public is granted the right to participate. Early case decisions which involved the standing of industrial plaintiffs left such standing in doubt, because the courts apparently are not as yet convinced that NEPA was designed to protect corporate interest in the environment in the same manner in which Congress intended to protect the public's interest. Public intervention adds a new facet to responsive and responsible decision making, while establishing a safety valve whereby the opinions and views of the public can be presented before an action is approved and implemented.

Environmental impact statement preparation cannot adequately fulfill the goals of NEPA if the impact assessment is not integrated into the entire planning process. Since the planning process usually involves intuitive judgments, predictions, and assumptions concerning alternate plans, the impact assessment must be a continuous activity if it is to influence the planning at all stages, as intended by NEPA.

The impact assessment cannot be considered to be a separate function which occurs after the major problem areas have been defined and the choices narrowed to one or two alternatives. The impact statement process has already exerted beneficial effects on the planning process. The critical review, which an agency knows the EIS will receive from the public, often is an incentive to the agencies not to propose projects which have known detrimental effects. The language of NEPA makes it clear that it is intended to be more than simply an environmental full-disclosure law. Congress intended NEPA to effect substantive changes in decision making.

Since the draft impact statement must be released to the public, as well as to concerned federal agencies, it seems obvious that comments on the statement will be received from environmental groups and interested citizens. Since the final impact statement must then respond to these comments, as well as to the comments of governmental agencies, environmental harm which may have been overlooked is brought to the attention of the agency proposing the action. Before filing the final EIS, agency personnel *must* make proper response to any comment which is relevant and reasonable. Also, the filing agency is required to conduct any necessary additional research to provide satisfactory answers, or to refer to those places in the draft statement which address themselves to the question raised where the agency provides adequate answers. If the final impact statement fails substantially to adequately answer all questions, comments, and criticism raised in regard to the draft statement, the final statement will not meet the requirements of the law.

4-8. SUMMARY

When Congress passed the National Environmental Policy Act in 1969 it was quite clear in its intention that the Act should include a statement of the nation's environmental policy. In addition to this statement of policy, it was further the intention of Congress that some type of action-forcing provision should be included in the Act, which would require all agencies of the federal government to comply with the requirements as detailed. The heart of this action-forcing provision is the Environmental Impact Statement. Federal lawmakers, as well as federal judges, have been quite specific in their requirements that each federal agency assess in depth the probable impacts of any action proposed. In the words of the law, when such action is major and federal or if it is one which will significantly affect the environment, the agency is required to provide for public input at an early stage in their planning. On the basis of the predicted environmental impacts, and at later stages of the planning and implementation of the project, public participation and comment must again be solicited. Many court decisions have been handed down specifying various essential details of the draft statement, as well as the requirements for the final impact statement. Such topics as when an action is sufficiently significant to require an impact statement, what must go into the statement if one is required, and what is the relationship of the agency to the public and other federal agencies, have all been addressed in the landmark court decisions. The minimum requirements that must be satisfied if no statement is to be issued have been outlined both in court decisions and in the guidelines handed down by the Council on Environmental Quality, as well as by the Environmental Protection Agency.

The draft environmental impact statement must be prepared by the agency proposing the action. This agency may contract with outside consultants, a private firm, or even with another federal agency to make the assessment of the predicted impacts; however, the actual statement must be prepared, as specified in the law, by "a responsible federal official" of the agency proposing the action. The draft statement must be circulated to the EPA and to all other agencies and branches of local, state, and federal government having expertise in the areas to be affected by the proposed action.

This statement is required to be circulated sufficiently early to allow all of the agencies, interested public groups, and individuals adequate time to comment on the draft statement. In addition to requiring the allotment of sufficient time for external comment, the preparing agency must include the comments as received, along with sufficient explanation or answers to the comment, in the final impact statement which is ultimately filed with CEQ.

There are two specific requirements to which the courts have attached the utmost importance, that must be addressed in any statement. The first of these is the no-action or "null" alternate. This alternate of doing nothing instead of the proposed agency action must always be considered in the statement as circulated. A second requirement, on which Congress and the courts are quite specific, is that a reasonable number of alternate solutions to the proposed action must be thoroughly investigated and their impacts included in the draft statement. Not only must the proposing agency specify a reasonable number of alternates to the full project, but it must specify, where possible, alternate solutions to specific parts of the proposed action. This is particularly true in those cases where a substitution of an alternate for a particular part of the project will result in a less severe environmental impact.

In order that no potential impacts are overlooked, CEQ, in one of its initial requirements after the passage of NEPA, has specified that each federal agency must prepare and issue a set of specific guidelines, detailing the method by which they will prepare statements or assess the impacts of any proposed action. These guidelines, as circulated by the various federal agencies, have been carefully considered by both the CEQ and the courts. The courts, in several instances, have ruled that a draft environmental impact statement as submitted was inadequate because an agency had not completely fulfilled the conditions of its own guidelines. Examples of several typical agency guidelines, as well as several typical tables of contents of completed EIS statements, have been included in this chapter.

Perhaps the most important item which must be kept in mind by agency personnel, or by a contracted external consultant who is preparing a draft statement or assessing the probable impacts of an action, is that both the courts and Congress have intended the entire procedure to be one of fish-bowl planning. It is the intention that all proposed actions and any resulting impacts be fully discussed with the public, in language that is easily understood by the layman, and early enough in the project planning that, without undue delay of the project or additional unbearable expense, changes can be made when appropriately called to the planning agency's attention. Thus, at all stages of the project planning and implementation, great care must be taken to ensure that the public is fully informed and aware of any changes in the environment or of irretrievable commitments of resources that will occur as a result of any agency action.

REFERENCES

4-1. The National Environmental Policy Act of 1969, Public Law 91-191, 91st Congress, S 1075, January 1, 1970.

4-2. Environment Defense Fund, Inc. v. Corps of Engineers (Gillham Dam), 325 F. Supp. 728, 1ELR 20130 (E.D. Ark.) (1970–1971).

4-3. *Hanly v. Kleindienst,* F. Supp. 3ELR 20016 (S.D.N.Y.) (Hanly II) (1972).

4-4. *Texas Committee on Natural Resources v. United States, 430* F. Supp. *1315,* 2ELRr20574 (W.D. Tex. 1970).

4-5. *Environmental Defense Fund, Inc. v. Environmental Protection Agency, 331* F 2d *925,* 6ERC 1112, (D.C.C.R 1973).

4-6. *McQueary v. Laird,* 459 F. 2nd 608, 1ELR 20607 (10th Cir. 1971).

4-7. *Nielson v. Seaborg,* 348 F. Supp. 1369, 2ELR 20765 (D. Utah 1972).

4-8. *Calvert Cliffs' Coordinating Committee v. Atomic Energy Commission,* 449 F. 2nd 1109, 1ELR 20346 (D.C. Cir. 1971).

4-9. *Hanly v. Mitchell, 460* F. Supp. *640,* 2ELR 20181 (S.D.N.Y. 1972).

4-10. *Scherr v. Volpe,* 336 F. Supp. 882, 2ELR 20068 (W.D. Wisc. 1971).

4-11. *Citizens for Clean Air v. Corps of Engineers,* 349, F. Supp. 696, 2ELR 20650 (S.D.N.Y. 1972).

4-12. *Lathan v. Volpe,* 445 F. 2d 1111, 1ELR 20602 (9th Cir. 1971).

4-13. *Sierra Club v. Hardin,* 325 F. Supp. 99, 1ELR 20161 (D. Alas. 1971).

4-14. *National Resources Defense Council v. Grant,* 341 F. Supp. 356, 2ELR 20185 (E.D.N. Car. 1972).

4-15. *Environmental Impact Statement Guidelines,* U.S. Environmental Protection Agency, Region X, Seattle, Washington, Revised Edition, April 1973.

4-16. *Environmental Defense Fund v. Corps of Engineers* (Tennessee-Tombigbee Waterway) 331 F. Supp. 925, 1ELR 20466 (D.D.C. 1971).

4-17. *Natural Resources Defense Council v. Morton,* 337 F. Supp. 165, 2ELR 20028 (D.D.C. 1971).

4-18. *Environmental Impact Assessment: Guidelines and Commentary,* Thomas G. Dickert and Katherine R. Domeny, University of California, Berkeley, California, 1974.

5

Assessment Methodology

5-1. INTRODUCTION

In the first four chapters of this book, the background to the present program of environmental assessment was given. Also, the legal background in common law and in statutory law for environmental litigation was established. In Chapters 3 and 4, the background and provisions of the National Environmental Policy Act of 1969 were described. These chapters have served to set the stage for a presentation of the methods and techniques used in actual assessments of environmental quality and in predictions of impact of proposed activities on environmental quality. This chapter is devoted to such a description. In the first part of the chapter, the methods available for use in environmental assessment are described. An approach is suggested for use in assessment operation. In the second portion of the chapter the organization of personnel for environmental assessments is described.

In order to illustrate the techniques that are described in the following sections of this chapter, the design and construction of power plants has been selected as a type of activity which is subject to considerable investigation with respect to environmental impact. It is appropriate to discuss impact assessments with particular respect to power plants, because of the ever-increasing demand for power in industrial societies and developing countries. In areas which are experiencing rapid economic growth, very great pressures will be exerted for expansion of power generation facilities. In response to such pressures, siting and construction of power plants may be carried out without a thorough analysis of the effects on the surrounding environment of plant construction and operation. The development plan for any type of power generation facility should include a thorough examination of the possible effects of the plant so that the developers can be sure that power plant construction and operation will not degrade the quality of the environment surrounding the plant. Obviously, power plant planning should include a comprehensive environmental assessment.

It is obvious that power plants will be built in ever-increasing numbers throughout the world in response to the great demand for power. These plants can have significant effects on the environment and therefore should be examined in light of their possible environmental impact. Fossil fuel power plants can have a significant effect on the environment, especially through the generation of particulate matter and sulfur dioxide which are released to the atmosphere. Nuclear powered generating facilities may create hazards because of the radioactive nature of the energy source, and because of the radioactive nature of the waste materials produced in the power plant. Also, both fossil fuel and nuclear-powered plants exert a great demand on surrounding water supplies for cooling water (Ref. 5-23). Subsequent release of cooling water into streams can lead to serious detrimental effects on aquatic life because of the elevation in temperature of the water during its passage through a power plant.

Power plant siting, construction, and operation also can lead to misuse of land through improper disposal of the wastes generated in the plant. Improper disposal of nuclear wastes can produce contaminated food sources and water supplies. Ash produced in coal-fired power plants may not amount to a very great quantity of material (Ref. 5-36), but large quantities of wastes are produced at the coal mines where the fuel for the power plant is obtained. Thus, in considering wastes produced in power generation, both fuel processing wastes and mining wastes should be included. In addition to these wastes, solid waste materials from power plants may be created through air pollution control systems which remove particulate matter and sulfur dioxides from stack gases. These pollutant substances originally present in fossil fuels are released to the atmosphere in most conventional power plants. At the present time, much of the particulate matter generated in power plant boilers is removed from stack gases through the use of electrostatic precipitators. In the near future, a significant effort will be expended to install additional apparatus on power plants to remove more particulate matter and to remove sulfur dioxides from stack gases through wet scrubbing techniques. The effluent from these wet scrubbers will constitute yet another waste material resulting from power generation. Disposal of these sludges will constitute a serious demand for land adjacent to power plants, and may pose a serious environmental threat if toxic materials from the coal fuel are collected in the scrubbers and find their way into the sludge. If leachate produced from the sludges migrates into surface waters or groundwaters, a very serious problem will be created (Refs. 5-16, 5-18, 5-36).

The possible impact on the environment of fuel processing and pollutant collection must be considered with any possible hazards which may result from power plant accidents. These effects must be included in any study of possible location, design, and operating characteristics for proposed power plants. The assessment of environmental effects of power generation is required by the provisions of the National Environmental Policy Act of 1969, and by subsequent Executive Orders (Ref. 5-40). Alternative locations for proposed power plants must be examined, as well as alternate fuels and alternate modes of operation.

From the description in the preceding paragraphs, it should be apparent that the siting, construction, and operation of power plants is a very pertinent activity with respect to environmental assessment activities. Consequently, this example will be used further in subsequent paragraphs to illustrate techniques of assessment. Other activities which could be subjected to comprehensive assessment and evaluation are described in terms of case histories in Chapter 6. It should not be supposed from the foregoing remarks that power plant location and design has been undertaken in a haphazard fashion in the past in the United States. Rather, it is more appropriate to say that power plant location, design, and operation have been governed by technical and economic considerations, rather than by environmental considerations. For example, the "best" location of a power plant in a given region has been based in the past upon the physical characteristics of the region and the ability of the earth materials at a given site to support the loads of the physical plant. Additionally, the technical alternatives for disposal of mining wastes and combustion wastes from power plants have been major factors in the location and design of such facilities. In this connection, natural geologic processes have played an important part in the location of power plants. In nuclear plant operation, for example, the most important factors to be considered include the requirement for human safety with respect to radioactive material, and the requirement to consider disturbance of a plant at a particular site because of geologic processes. The most serious threat to nuclear power plants is the major disturbance associated with an earthquake. Failure of the plant containment vessel, with consequent release of radioactive material, represents an extreme hazard. Imposition of earthquake forces on the components of a nuclear power plant could cause such failure. Therefore, power plant site selection must include a consideration of seismic activity, although earthquakes actually represent random events in a continuing geologic process. In response to this need to consider earthquake forces in the design and construction of power plants, design criteria have been developed to safeguard nuclear power plants against the effects of shock waves produced during earthquakes. These criteria also are intended to protect nuclear plants from the possible changes in the characteristics of foundation materials which could result from the imposition of dynamic earthquake forces (Ref. 5-24). In the United States, regulatory agencies have described and defined the earthquake for which nuclear power plants must be designed (Refs. 5-31, 5-33). It can be seen that in the past, geologic and seismic criteria have been the predominant factors in evaluating sites for nuclear power plants (Ref. 5-32).

Construction and operation of fossil fuel power plants do not include the potential hazards of radioactivity associated with nuclear power plants. Consequently, location of fossil fuel power plants has been based almost exclusively on economic considerations. Additionally, some attention has been given to the problem of waste disposal from fuel combustion, and to the effects of air and water pollutants emitted by such plants. In the early days of applications of electrical power in industry, power plants were located near the centers of demand areas. As cities grew and land costs increased dramatically, power plants were moved from the centers of urban areas to suburban locations. As citizen resistance to the side effects of power generation (particularly air pollution) increased in recent years, power plants were moved farther and farther away from cities. Also, as the size of power plants increased, the need for cooling water and the capability to safely discharge large volumes of heated water also increased. It became necessary to create large reservoirs of water for use in cooling power plants, and remote sites were needed for the construction of such reservoirs. Consequently, power plants were removed farther and farther from urban areas. In this trend of changing locations for power plants, environmental factors played little part, with the exception of the citizen resistance to air pollution associated with fossil fuel plants. However, in recent years, changes in power plant operation have made it possible to consider locating power plants much closer to urban centers. Air pollution control systems have been improved, and cooling water systems are being developed which will allow closed-loop operation. It now appears feasible to return power plants to the fringes of urban areas in order to reduce transmission losses and consequent costs. If power plants are moved close to population centers, the problems of waste disposal and environmental hazards to individuals will become even more intense. Also, the effects of construction operations will become pronounced, especially the social and economic effects of such construction (Ref. 5-38). In the case of developing nations, the impact of power plant construction and operation on the *physical* environment may be of greater concern to citizens than local, social considerations. However, long-term considerations may indicate greater importance for such local political and social impacts. Therefore, it seems quite clear that location, construction, and operation of power plants in the future should be subject to an evaluation which includes a comprehensive assessment of environmental effects. Power plants are but one class of facilities which will require such assessment. They merely serve here as a useful example.

The preceding paragraphs were intended to indicate, for the example of power generation facilities, the need for comprehensive environmental assessment. Such assessments are extremely important in evaluating proposed plans for the location, construction and operation of power generating facilities. Many other classes of activities and projects could be listed as requiring comprehensive environmental assessment, in addition to the power plants which have been used as examples. It should be obvious, however, that regardless of the example used, an important need exists for comprehensive methods of systematic environmental assessment (Refs. 5-7, 5-18). The authors have participated in a number of studies of environmental impact for several different types of proposed facilities. During the course of these studies, they have developed some conclusions concerning methods of assessment and methods of personnel organization (Refs. 5-9, 5-10, 5-11, 5-12). These conclusions are presented in this chapter in the hope that they will be of assistance to other investigators, officials, and citizens who are involved in performing environmental assessments.

5-2. AVAILABLE METHODS

It is important to review all methodologies which can be applied in performing environmental assessments because of the inherent importance of the assessment process. Also, a comprehensive and systematic assessment methodology will lead to mitigation of impact and will lead to profitable alteration and re-design of facilities, as well as suspension of plans for undesirable activities. For the assessment of power plants as described in previous paragraphs, the importance of examining alternative sites, alternative fuels, and alternative modes of power generation lies in the possibility of modifying a proposed plant site, plant design, or operating procedure, in order to minimize undesirable environmental impact during construction and operation of the proposed plant. If a comprehensive analysis of environmental impact is completed for a given facility, proposed designs and plans can be modified to minimize detrimental impacts and to maximize benefits from proposed activity.

5-2.1 Traditional Methods

It is much easier to describe the need for comprehensive assessments and to describe the benefits of thorough evaluation of alternatives than it is to perform such assessments and evaluations. Environmental impact identification and evaluation is very complex. Several attempts have been made in the recent past in the United States and other countries to develop standard techniques and methods for environmental assessment. These methods, which have been described in the technical and professional literature, have been reviewed. In general, no one method has been found to be adequate for all situations. Traditional methods are directed toward establishing technical and economic values for alternatives. For example, equivalent uniform annual cost, present worth, net present value, rate of return, and benefit/cost ratio analyses as well as other techniques have been applied in the past by engineers and managers in evaluating proposed activities. These techniques are not adequate for environmental assessment because it is impossible to establish economic values or monetary worth for many environmental characteristics. For this reason, many of the traditional methods which have been used in the past to assess alternative plans and designs are not applicable in environmental assessment.

5-2.2 Graphical Techniques

In addition to traditional economic evaluation techniques, many traditional methods of graphical representation now are being used to portray the location and severity of impacts on the environment, as well as the location and quality of valuable segments of the environment such as marshlands (which are particularly productive in a biological sense). Graphical comparison techniques have been applied to land-use studies and environmental assessment investigations by a number of individuals and agencies. One of the most thorough applications of graphical representation techniques has been made in the use of maps to portray areas of impact and areas of particular value. Perhaps the foremost proponent of this type of activity has been Ian McHarg. McHarg's techniques are presented in his book, *Design with Nature* (Ref. 5-22). In this book, the author gives numerous examples of the use of graphical aids in evaluating environmental quality and the limitations to use of land areas and resources. For example, he presents comprehensive data on Staten Island, a unique resource in the metropolitan area of New York City. McHarg uses the Island as an example for application of his technique. Essentially, he portrays, by means of maps, the basic physical condition of the Island and the location of unique or valuable features. He contends that any locality is a product of the total sum of biological, physical, and historical processes that have shaped the locality. He also contends that these processes are ever-changing and that the inherent social value of a particular locality is governed by, and expressed in, these changing processes. The basic question for the Island in the example presented by McHarg is the conflict in proposed land uses. What portions of the Island should remain untouched? What areas should be developed for residential sections? What recreational areas should be established and/or preserved?

McHarg presents a description of the Island in terms of the bedrock geology, the surficial geology, the hydrology, and the basic soil drainage environments of this unique landform. He supplies text to accompany maps which show the geographic location of various geologic and hydrologic features. His contention is that the social value of a particular terrain is dependent upon the suitability of that terrain for a variety of uses. For example, flat-lying well-drained land may be used for commercial or industrial development, but it also may be used for residential development or recreational purposes. On the other hand, certain areas with abrupt changes in topography and poor drainage characteristics may not be suitable for commercial, industrial, or residential development, but may possess high scenic value and great recreational potential.

As mentioned previously, the physical characteristics of the Island are presented by McHarg in the form of maps describing the basic landforms and water flow patterns in the area. An overall summary map showing physiographic features also is presented. Geologic cross-sections of the Island are included in his description, as are maps showing unique geologic features, areas inundated by tides, areas of steep slopes, areas where development is limited by poor foundation conditions, and areas where development is limited by the presence of water at a shallow depth beneath the surface. Ecological features also are portrayed in maps of the Island. For example, existing wildlife habitat, existing vegetation, existing qualities of forested land, and existing ecological associations in forested areas are shown on maps. The effects of man's use of the Island are portrayed in maps showing historical landmarks and existing land uses for the entire Island.

Each of the maps mentioned above is prepared through using a variety of colors or tints to portray varying qualities or different conditions. For example, the areas on the Island where development is limited by poor foundation conditions are portrayed in gray, black, and green colors with the different colors corresponding to different degrees of limitation. Obviously, the degree of limitation for a particular use may be different from the degree of limitation for a dissimilar use. For example, in selecting areas which are suitable for conservation, McHarg listed factors which were of greatest importance: features of historic value, high quality forests and marshes, bay beaches, streams, water-associated wildlife habitat, intertidal wildlife habitat, unique geological and physiographic features, scenic land and water features, and scarce ecological associations. For residential land use, factors which would favor such use include attractive surroundings. Thus, scenic land features, locations near water, and the presence of historic sites and buildings in the immediate area would favor residential development. On the other hand, steep slopes, susceptibility to flooding and poor drainage conditions would be factors which would limit residential development. In his book, McHarg included maps which show the degree of quality for these features (historic features value, beach quality, existing forests quality, intertidal habitat value, etc.) for each of the particular uses under investigation. In general, the highest quality of any particular attribute, such as beach quality, would be shown in terms of the lightest color in a given range or as a greater intensity of a particular shading. The maps which portrayed these differences in quality were prepared on transparent film. Thus, the maps could be superimposed to give a composite picture, where intensity of shading or intensity of color would indicate a cumulative evaluation, for a particular use, of all pertinent factors. For example, conservation areas were determined by a superposition of maps showing the quality of features of historic value, forest areas, marsh areas, bay beaches, streams, and many other scenic features. In evaluating the recreation potential of various areas on the Island, McHarg proposed the use of similar features with the inclusion of existing land uses and potential recreational habitat. The potential for urbanization of the Island was evaluated in similar fashion. Conservation areas, recreation areas, and urbanization areas were represented on transparent maps in different color codings: conservation areas were shown by means of various shadings of green; recreation areas were represented in blue tones; and areas for urbanization were evaluated in terms of gray, dark gray, and black. The overall evaluation of the Island for future use was portrayed vividly and dramatically by a superposition of the three maps, representing the potential of the area for conservation, recreation, and urbanization.

The techniques developed by Ian McHarg are among the most useful graphical comparison techniques which have been developed for use in land-use planning and environmental assessment. Other graphical comparison techniques include computer map analysis, map overlays (Ref. 5-22), shaded matrices (Ref. 5-4), value profiles (Ref. 5-25), relative merit analysis (Ref. 5-1), physical models (Ref. 5-12), and contour mapping of impacts (Ref. 5-26). In general, graphical comparison techniques have been found to be most appropriate as communication tools rather than as methods of assessment. These methods are useful in communicating to the public the nature and degree of particular impacts. Although these methods do not allow quantitative comparison among alternatives, they do elicit valuable input from non-technical personnel, especially the public. The more recent developments in the use of interactive, computer-based graphics are particularly valuable (Ref. 5-30).

Interactive graphics have been found to be extremely useful in several ways in communicating with the public. For example, these graphic aids have been particularly useful in generating new ideas and in soliciting suggestions from the public concerning the relationship between existing communities and proposed activities, and the consequent impacts thereof. The ability to graphically portray, in a short time (especially for computer-aided systems), the effects of proposed actions and alternatives has proven to be extremely valuable. These techniques also have been found to be useful in identifying potential problem areas as well as in finding solutions to those problems. In many cases, finding solutions to the problems consists of considering an expanded set of alternatives, and graphical techniques are extremely useful in identifying such alternatives. Of course, computer-aided interactive graphics can be used in establishing values and in evaluating trade-offs, as proposed by Mr. McHarg, and described above. The interactive graphics devices may consist of maps or such materials as bar charts, line graphs, tables, perspective drawings, pie charts, and similar devices. These devices can be used to illustrate the proposed design and possible alternative designs. One of their most positive advantages is the rapid portrayal of changes or alterations in design.

Interactive graphics can be used in the analysis of particular projects or in the study of entire systems. In both cases, graphic aids produced in this fashion are particularly useful in clarifying, for the public, the product of the feasibility studies, economic studies, and design efforts undertaken by the planners and engineers who have initiated the activity under study. Additionally, interactive graphics have been found to be extremely useful, in the comparison of alternatives, through the comparison of differences only. In other words, a particular alternative may be represented on a map or a similar device by the impact, or areas impacted, peculiar to that alternative. The areas affected by another alternative, but unaffected by the remaining set of alternatives, would be represented by a second map or graphical display. The areas remaining blank on all of the maps representing all viable alternatives would be those areas either unaffected by any alternatives, or affected equally by all alternatives. Since the null alternative must be included in the analysis, virtually all portions of a given study area would appear on one of the maps portraying the effects of the various alternatives.

For purposes of *citizen participation* in environmental planning, computer graphics have been found to be useful in several areas (Ref. 5-2). These graphic tools are valuable in portraying complex proposed activities and in permitting value judgements concerning the effects of these various complex activities. Additionally, these tools are useful in establishing a clear understanding of the geographic distribution and distances involved in particular projects. Finally, they have been found to be extremely useful in illustrating the methods and tools used by project planners, and in demonstrating the technical assistance which is available from the project initiators and the assessment team. All too often, citizens interested in particular projects, or persons functioning as members of citizen advisory committees, remain ignorant of the technical assistance which could be obtained by a simple request directed to project initiators or members of an impact evaluation group.

5-2.3 Numerical Methods

A number of numerical techniques of impact assessment and analysis have been proposed. These methods have been intended to allow quantitative comparison of alternatives and to fill a need unsatisfied by traditional methods or graphical techniques. The numerical systems which have been developed are intended basically to furnish an estimate of environmental quality, and changes in environmental quality, in terms of number values for selected parameters. The numerical assessment methods which have been developed include ranking systems (Ref. 5-4), value matrices (Ref. 5-27), expected value methods (Ref. 5-28), utility theory applications (Ref. 5-4), rating systems (Ref. 5-8), decision models (Ref. 5-15), and several other techniques. These methods have been reviewed in detail in other publications (Refs. 5-4, 5-7, 5-18). In general, simple numerical techniques are not valuable in environmental assessment efforts because of the difficulty, previously cited, in placing a numerical value, in terms of any given parameter, on environmental attributes. Most of the methods which have been proposed are extremely simple, to the point of being oversimplified. In general, they are not sufficiently discriminatory to allow proper evaluation of the impact from various alternative proposals. Some of the numerical methods which have been proposed have been refined to the point where they are useful for certain types of projects, or for certain segments of the entire assessment tasks. Because some of these methods have some utility, they will be described in subsequent sections.

5-2.4 Matrix Systems

As an example of a system which is useful to a certain extent in environment assessment, the technique developed by the personnel of the United States Geological Survey (Ref. 5-21) can be described here. This technique is particularly useful for identifying and categorizing proposed activities and for identifying anticipated impacts. This type of method is best applied in preliminary assessments of environmental effects. The system is based upon the use of a matrix for identifying possible problem areas, for representing possible impacts graphically to design professionals, agency personnel, and interested citizens.

According to the procedure recommended by the developers of the matrix, the system yields a quantitative estimate of impact; this estimate, however, may not give proper weight to the various aspects of a particular project in a specific locality. One hundred possible activities are listed in the matrix along one axis, with 88 possible impacts listed along the other axis. Thus, 8,800 interactions can be displayed or indicated by means of this matrix. Although a great number of possible interactions may be identified, only first-order direct impacts would be identified, with no more than 10 to 12 categories of impact considered significant for any one proposed activity, according to the developers of the matrix (Ref. 5-21). Figure 5-1 shows a portion of the matrix developed for environmental assessment activities by the United States Geological Survey personnel.

II PROPOSED ACTIONS

A. MODIFICATION OF REGIME

a. Exotic flora or fauna introduction
b. Biological controls
c. Modification of habitat
d. Alteration of ground cover
e. Alteration of ground water hydrology
f. Alteration of drainage
g. River control and flow modification
h. Canalization
i. Irrigation
j. Weather modification
k. Burning
l. Surface or paving
m. Noise and vibration

B. LAND TRANSFORMATION AND CONSTRUCTION

a. Urbanization
b. Industrial sites and buildings
c. Airports
d. Highways and bridges
e. Roads and trails
f. Railroads
g. Cables and lifts
h. Transmission lines, pipelines and corridors
i. Barriers including fencing
j. Channel dredging and straightening
k. Channel revetments
l. Canals
m. Dams and impoundments
n. Piers, seawalls, marinas, and sea terminals
o. Offshore structures
p. Recreational structures
q. Blasting and drilling
r. Cut and fill
s. Tunnels and underground structures
a. Blasting and drilling

INSTRUCTIONS

1- Identify all actions (located across the top of the matrix) that are part of the proposed project.

2- Under each of the proposed actions, place a slash at the intersection with each item on the side of the matrix if an impact is possible.

3- Having completed the matrix, in the upper left-hand corner of each box with a slash, place a number from 1 to 10 which indicates the MAGNITUDE of the possible impact; 10 represents the greatest magnitude of impact and 1, the least, (no zeroes). Before each number place a + if the impact would be beneficial. In the lower right-hand corner of the box place a number from 1 to 10 which indicates the IMPORTANCE of the possible impact (e.g. regional vs. local); 10 represents the greatest importance and 1, the least (no zeroes).

4- The text which accompanies the matrix should be a discussion of the significant impacts, those columns and rows marked with large numbers of boxes and individual boxes with the larger numbers.

SAMPLE MATRIX

	a	b	c	d	e
a		2/1		8/5	
b	2/1	8/3	9/7		

PROPOSED ACTIONS

HEMICAL CHARACTERISTICS		
1. EARTH	a.	Mineral resources
	b.	Construction material
	c.	Soils
	d.	Land form
	e.	Force fields and background radiation
	f.	Unique physical features
2. WATER	a.	Surface
	b.	Ocean
	c.	Underground
	d.	Quality
	e.	Temperature
	f.	Recharge
	g.	Snow, ice, and permafrost
3.	a.	Quality (gases, particulates)

Fig. 5-1 A portion of the matrix developed for environmental assessment by the U.S. Geological Survey (Ref. 5-21).

Each element (box) in the matrix consists of an upper portion in which the impact evaluator indicates the degree of impact of a particular action. The degree of impact is indicated numerically by a value from 1 (minimum impact) to 10 (greatest impact). In the lower portion of each box, the impact evaluator indicates the overall importance of the particular impact identified. This indication is done in a similar numerical fashion; i.e., the importance of the impact is rated on a scale from 1 to 10. The number rating for impact and importance of impact (the upper and lower numbers in each box of the matrix) thus are determined on the basis of value judgements made by the investigator. Obviously, this approach is a highly subjective technique, since no criteria are listed in the procedure, as developed, for making such judgements or for assigning numbers to the degree of impact and the importance of the identified impact.

The matrix approach is relatively inflexible. It contains no provision for considering secondary effects of a proposed activity. It would be possible to construct a matrix designed for the evaluation of secondary effects, but developing a simple correlation technique to relate the findings of the primary matrix review with the results from the secondary matrix review would be extremely difficult. Lack of consideration of secondary effects is a very serious defect in the USGS matrix system of impact assessment. Also, short-term and long-term studies would require separate matrices and would entail additional problems in communication. For these reasons, matrix methods are considered useful primarily in preliminary studies, for the identification of impact areas and for the designation of particularly fragile or controversial environmental attributes.

5-2.5 Checklist Systems

The matrix systems described in the previous section have been found to be useful as memory aids in the assessment process. It is extremely useful to have a comprehensive listing of possible impact areas against which an investigator can check the conditions of a project under study. This advantage has been increased in some methods of impact assessment which rely heavily on comprehensive checklists for identification of impact. In this type of system, an extensive listing of possible areas of impact on all segments of the environment is utilized. The project under study is compared, for possible impacts, with the listing of impacts and associated activities contained in the checklist (Refs. 5-20, 5-32, 5-34, 5-35, 5-37). The most significant advantage of the checklist method is the capability it gives to the investigator to uncover almost all possible areas of impact; it is far too easy for particular areas of impact to be overlooked, even by a very conscientious investigator. However, in balance, this method is useful primarily as a memory aid and not as a quantitative evaluation tool.

The checklist approach, in and of itself, is not adequate for a comprehensive environmental assessment. However, some checklist techniques have worthwhile characteristics. For example, personnel of the United States Atomic Energy Commission (AEC) have developed guidelines for environmental assessment which are rather comprehensive in that they require a number of descriptive evaluations of the proposed power plant. These descriptive evaluations include study of the plant purpose, site character, plant character, environmental effect of construction, impact of normal operation, effluent control and monitoring, effects of accidents, socio-economic effects, alternative energy sources and sites, plant design alternatives, summary benefit/cost analyses, lists of environmental approvals from and consultations with other agencies, and listing of documentary references, for any proposed power plant. This compilation of evaluations constitutes a very comprehensive checklist. It tends to be somewhat impractical, however, since it is very difficult to quantify all of the environmental attributes in a given area, affected by a specific plant, in such items as a benefit/cost ratio. Many environmental factors may not be expressed easily in terms of quantifiable costs or benefits.

One of the disadvantages of the use of a checklist method of environmental assessment is the fact that many such checklists produce ungainly tabulations of baseline data. These tabulations may be of little use to the investigators or to the general public unless some selective criteria are applied to the baseline data in order to pinpoint significant areas of impact. Selective criteria have been developed in some systems. For example, the guidelines developed by the United States Department of Housing and Urban Development (HUD) (Ref. 5-35) include the concept of "threshold" values for environmental parameters, and for changes in environmental quality expressed in terms of these parameters. These parameters and the associated threshold values can be utilized in the compilation of impact assessments. However, the HUD guidelines are a far cry from the AEC guidelines; they are not sufficiently comprehensive or systematic.

One of the most comprehensive and useful checklist methods of environmental assessment developed to date is that compiled at the U.S. Army Construction Engineering Research Laboratory in Champaign, Illinois (Ref. 5-20). This method has been described in considerable detail in *Environmental Impact Analysis: A New Tool in Decision Making,* by Urban, Jain, and Stacey. This valuable guide to the methodology of impact assessment is a companion volume in the Van Nostrand Reinhold Environmental Engineering Series.

In this method, the developers concentrated on the environmental impact of activities of the United States Army. However, the rationale and logic of the method may be applied equally to the activities of any agency or individual. The identification of possible areas of impact was facilitated in this method through the use of a matrix. The matrix consists of a listing of Army activities, and various environmental attributes which could be affected by those activities. Included in the activities are construction, operation, maintenance, repair, training, mission change, real estate changes, procurement, industrial activities, research, development, tests, evaluation, administration, and support activities. The environmental attributes listed in the matrix include air, water, land, ecology, sound, and socio-economic characteristics of the environment. According to the procedure proposed by the developers of the system, the assessment team preparing the impact assessment or study would define the proposed activity in as much detail as is suitable for the importance of the project. The various environmental attributes then would be examined, one by one, for possible impact. To aid the evaluation team in this examination, detailed descriptions of the various environmental attributes were prepared by the developers of the method. For example, under the general category of air quality, nine different particular attributes were established. These attributes included diffusion factor, particulates, sulfur oxides, hydrocarbons, nitrogen oxides, carbon monoxide, photo-chemical oxidants, hazardous toxicants, and odor. In comparison, for the general category of the land environment, three particular attributes were developed: erosion, natural hazard, and land use patterns. A total of 46 particular attributes were established for the eight general categories of environmental quality. Each of the 46 particular attributes was described in detail in a supplementary manual for the use of the assessment team. The characteristics of the particular attribute were given, and methods for determining the impact of a particular activity on a given attribute were listed. Each "attribute descriptor package" was presented in a standardized order and format in order to facilitate use of the information gathered in preparing an assessment. In general, in each package, the environmental attribute was defined, and an explanation was given as to how the particular attribute relates to the total quality of the environment. Examples of Army activities which could affect the particular attribute then were given and the sources of the predicted effect were listed. The real world variables that could be measured as indicators of changes in the particular attributes were also given in the package. Detailed instructions on the measuring techniques, the equipment needed, and the types of skills needed in the monitoring personnel were given. Also, instructions were given for evaluation and interpretation of the collected data. Special conditions for measuring certain attributes, geographical and temperal limitations on measurements, and means to mitigate impact in certain areas also were listed in each descriptor package.

The developers of this method indicated that the results of using their checklist technique could be represented in a variety of ways. They mention the fact that numbers could be assigned to each attribute and the degree of impact on each attribute could be indicated numerically, but they did not prefer this technique. Although this approach does provide a single number to indicate the total environmental impact of each alternative, the single number indicating total impact does not indicate the distribution of impact among the various attributes of environmental quality. Impact may be

Per capita consumption
Public sector revenue
Regional economic stability
Community needs
Physiological systems
Psychological needs
Life styles
Social behavior effects
Performance effects
Communication effects
Psychological effects
Physiological effects
Aquatic plants
Natural land vegetation
Threatened species
Field crops
Fish, shell fish, and water fowl
Small game
Predatory birds
Large animals (wild and domestic)
Land use patterns
Natural hazard
Erosion
Fecal coliform
Aquatic life
Toxic compounds
Nutrients
Dissolved solids
Dissolved oxygen (DO)
Biochemical oxygen demand
Acid and alkali
Thermal pollution
Suspended solids
Radioactivity
Oil
Flow variations
Aquifer safe yield
Odor
Hazardous toxicants
Photochemical oxidants
Carbon monoxide
Nitrogen oxide
Hydrocarbons
Sulphur oxides
Particulates
Diffusion factor

Socioeconomic — Human, Economic
Sound
Ecology
Land
Water
Air

ATTRIBUTE NUMBER: 1 2 3 4 5 6 7 8 9 10 11 12 13 14 15 16 17 18 19 20 21 22 23 24 25 26 27 28 29 30 31 32 33 34 35 36 37 38 39 40 41 42 43 44 45 46

*Net Positive Impact +

Net Negative Impacts X

Project Name _____

Project Number _____

Alternative _____

☐ No Significant Impact

▨ Moderate Impact

■ Significant Impact

*Positive impacts are shown above the attribute number and negative impacts below.

Fig. 5-2 Summary sheet for impacts developed by Jain *et al* for the Department of the Army (Ref. 5-20).

extremely severe in one particular area and very minor in other areas, but the resultant total rating would indicate rather minor impact on the total environment. The developers of the system indicated that the approach they would favor is a combined system, to indicate the severity of impact on all attributes in a single diagram, with an indication of the quality of the impact, whether positive or negative. To avoid any difficulties associated with controversy about the relative importance of various attributes, all attributes listed in this checklist were assumed to be of equal importance. The developers of the method suggested the use of a chart, similar to the one shown in Figure 5-2, as a graphical summary for presentation of the result of the environmental assessment. The severity of impact would be indicated through the use of shading in the diagram, and the quality of the impact (negative or positive) would be indicated by the location of the shading above or below the attribute number in the diagram.

5-2.6 Quantitative Systems

In addition to the systems of impact assessment described above, some systems have been developed to yield a high degree of quantitative evaluation, especially for projects of a particular nature, or for projects in a particular area. For example, a system of quantitative evaluation was developed by the personnel of the Battelle Memorial Laboratories for use in the study of alternative designs for water resources projects (Ref. 5-3). In this system, the method developers assigned numerical values to a long list of possible impact areas, by means of a complicated weighting procedure. By means of this complicated technique the developers of the system claimed that a quantitative evaluation could be obtained and that this system could be used to compare alternative locations for facilities and alternative modes of facility operation.

In the Battelle system, four basic areas of environmental quality are included for study: ecology, environmental pollution, aesthetics, and human interests. The Battelle investigators selected 78 parameters to characterize the environmental quality in these four broad categories. In order to give a numerical value to the environmental quality characterized by each of the parameters, a total of

1,000 parameter importance units (PIU) were distributed among the 78 parameters. According to the procedure suggested by the system developers, the "environmental quality" of each parameter could be estimated for a given project or proposed development, on the basis of a value of 0 for the worst possible conditions and a value of 1 for the best possible conditions. This environmental quality number (0-1) then would be multiplied by the value of parameter importance units (PIU) for each parameter, to yield the overall environmental importance of that parameter. This number was referred to as the number of environmental importance units (EIU) for that parameter. The environmental quality prior to the construction and operation of a particular project could be established by means of this system, as could the effects of any proposed activity. After the environmental importance units were obtained for each parameter, the sum of all the units for all the parameters could be obtained to indicate, quantitatively, the degree and character of the impact of the proposed action on the environment. Alternative plans could be evaluated in a similar fashion and a choice could be made among alternatives on the basis of the EIU scores.

In this procedure, certain environmental elements were singled out as especially important. These elements included politically sensitive issues and ecologically fragile elements. Although this attention was focused on particular issues, the system, as a whole, is inherently inflexible. The system furnishes numbers for purposes of comparison among alternatives, but it contains no provisions for identifying the parameters which are most important in a particular situation. In other words, the total scores of various alternatives would be used in evaluating the impact of those alternatives, although a given alternative could have very severe impact in one particular area of the environment. The severity of impact in a particular area could be a very restrictive factor in the implementation of the proposed activity, although the total score in the rating system would not indicate that that alternative was undesirable. Additionally, the system contains no provisions for giving extra emphasis to particular parameters by means of weighting factors associated with unique features in a particular area. Finally, the system was developed for use in evaluating water resources activities and is not strictly applicable to other forms of activity without some modification.

This system is typical of other quantitative methods of environmental assessment in that it contains the results of value judgements made by system developers. These value judgements may not be correct or pertinent to the particular project under study. Also, the quantitative estimate of environmental impact may lead to a false impression of precision in predicting impact. In other words, a multidigit number indicating a high degree of environmental benefit from a particular alternative may suggest to the uninformed reader that a high degree of accuracy could be attained in making the prediction of environmental impact. On the contrary, many approximations and guesses in estimating degree and type of impact could yield just as impressive a number in this type of system.

5-3. PROPOSED APPROACH

It should be obvious from the comments given in the preceding sections, that the authors feel that no one system of environmental assessment is applicable to all types of projects in all areas of the world. No one given system has overall general applicability, nor is any one system sufficiently flexible and comprehensive to cover every eventuality. However, certain general techniques can be utilized in any environmental assessment to avoid duplication and to achieve a rational and logical investigation of environmental quality and possible impact.

5-3.1 System Components

The approach suggested for environmental assessment would entail the use of a number of assessment tools in a definite sequence of events. One of the most useful tools in preliminary studies is a matrix similar to that developed by the United States Geological Survey or that developed by personnel of the U.S. Army Construction Engineering Research Laboratories. In conducting an assessment for a particular facility, these matrices could be used by the assessment team after they were modified to reflect the particular activities associated with the proposed project, and particularly important segments of the environment in the area of the proposed project. The matrix should be developed in a cooperative effort which would include participation by the project initiators, the assessment team, and selected members of the public. Mechanisms for insuring this cooperation among diverse groups are described in the later sections of this chapter.

After a matrix is developed as a preliminary analysis tool and as a means of communication among the various persons engaged in the assessment activity, mechanisms should be developed to include the public in the assessment process. Public hearings are a virtual necessity in this type of activity. Particular techniques to be used in organizing public participation are described in the next section. The analysis tools which are particularly useful in achieving public involvement include many of the graphical techniques listed in Section 5-2.2. Computer-generated graphic displays are particularly important in public hearings and in meetings designed to encourage public participation in environmental decision-making (Ref. 5-13). These graphics may be passive graphics or they may be interactive graphics. Passive graphics include maps, drawings, and various tabular displays which are prepared before public hearings and meetings and which do not permit interaction and input by the public. Land use plots, perspective views of particular facilities, and comprehensive evaluation maps, similar to those described in Section 5-2.2, could be included in the graphic material prepared before public meetings. In general, interactive computer graphics could include many of the same types of displays, but the parameters selected for the interactive displays should permit rapid change so that a computer could quickly create a new display. Computer assisted techniques are powerful in this regard. For example, computer-generated animation has been used successfully to simulate an observer's approach to a particular project or to simulate driving along a projected highway. Technology is available at the present time to permit use of this type of technique in public hearings and public meetings. This type of activity should be encouraged, since it is extremely useful in attracting citizen interest and in encouraging public input.

At the same time that preliminary plans are being displayed to the public in public hearings, some efforts should be devoted to developing a comprehensive checklist for use in assessing impacts from the proposed activity. In this development, public participation should be encouraged. One of the most useful techniques in this type of activity (developing a checklist) is to group classes of projects together so that general impacts can be identified easily. This could be done on a national basis as well as on an individual project basis. If the checklist were developed to indicate broad areas of impact from particular classes of activity, more particular attention could be given to specific facets of the environment in question and particular aspects of the proposed activity, rather than to duplication of general efforts. The checklist developed by the personnel of the U.S. Army Construction Engineering Research Laboratories could serve as an excellent beginning for the development of a checklist for a particular type of activity. In assigning importance to particular attributes of environmental quality, it is necessary to obtain citizen evaluation so that the true social value of various parameters could be established. Several techniques have been developed to obtain systematic citizen participation in environmental planning. One of the most useful techniques is a special voting technology for use in public meetings (Ref. 5-29). In this technique, various parameters and alternatives may be suggested for evaluation by a large group of citizens, design professionals, agency personnel, etc. By the use of a hand-held dial indicator, connected with a central computerized tally apparatus, overall evaluation (by all persons holding an evaluation device) could be summarized and displayed immediately in the public meetings. This type of technique, essentially, would permit secret balloting on the importance of various environmental parameters with regard to specific proposed activity. Quantitative statistical procedures could be used to evaluate responses, and alternatives could be compared through use of this type of technique.

During the development of an impact matrix and an assessment checklist, and during the progress of public meetings, great effort should be expended to simplify the environmental assessment being conducted. The method should be examined to see that it does include all pertinent principles of systems science. Additionally, in the past, physical characteristics of the environment have been over-emphasized, and more emphasis should be placed on analyzing utilization of resources, rather than on compilation of inventories and volumes of baseline data and physical environmental characteristics. More emphasis should be placed upon the prediction of future conditions, rather than on the characterization of the present conditions. First priority should be placed upon the development of criteria for the selection of environmental parameters of the greatest significance in the project area and for the particular type of project under study. In developing these criteria for selecting significant environmental parameters, consideration should be given to the frequency at which any given parameter is affected by the type of proposed activity, the seriousness of the impact possible from this type of activity on any given parameter, the severity of the indirect effects on this type of activity on that given parameter, and the degree of public controversy surrounding a given parameter.

5-3.2 Sequence of Events

The general sequence of events in completing an environmental assessment can be judged from the information given in the preceding section, from the guidelines on environmental assessment included in the National Environmental Policy Act, and from the recommendations concerning organization of personnel which are made in following sections of this chapter. However, at this point it is pertinent to emphasize that public involvement in the assessment process must be achieved at a very early date in the life of the project and must be continuous through the assessment process. In the method suggested here, a matrix should be developed, a checklist should be prepared, and parameters of particular importance should be identified. These operations can be undertaken in logical sequence by project initiators and by members of a professional assessment team. However, their work may be futile if the public is not involved in the entire process. Ensuring public involvement in the assessment process, however, is not an easy task. Some progress has been made in achieving effective citizen participation and some basic concepts have been developed (Ref. 5-19). In order to achieve effective citizen participation, viable and achievable objectives must be identified for citizen members of assessment teams. Additionally, specific and easily understood criteria must be established for the selection of citizens to participate in the assessment process. These criteria must be related to the objectives for the citizen participation and they must be related to the overall project objectives. Additionally, personnel of the initiating agency and members of the assessment team must develop close contact with citizen representatives on advisory committees and with influential citizens in the communities affected by the proposed project, in order to develop credibility for the assessment procedure.

In previous efforts to obtain citizen involvement in environmental assessments, the most successful programs were developed around certain basic concepts. The citizens should feel that they have something definite to contribute to the assessment process and that the assessment studies constitute an opportunity to present their views and recommendations. If the citizen reviewers of the proposed activity feel that their actions and comments are important, they will give appropriate consideration to their input in the assessment process. In developing criteria for selecting citizens to represent the general public on advisory committees and similar groups, an effort should be made to identify specific tasks for citizen evaluators so that the suitability of an individual to be an evaluator can be determined easily. Other general or specific criteria also could be developed for selecting citizen representatives; e.g., a citizen evaluator could be selected for membership on an advisory committee on the basis of experience with the type of project being proposed. Finally, the personnel associated with developing the project and with assessing the impact of the proposed activity must establish close and frank relations with members of the general public as well as with members of citizen advisory committees and panels.

Citizen participation in environmental assessment will differ according to the types of projects being proposed and the importance of the particular project in the total community (Ref. 5-6). The importance of citizen participation in the assessment study will govern the techniques which are used to recruit citizen members of advisory committees and review panels. The structure of advisory committees and even the structure of public meetings will be determined by the type of input which is necessary from the public. One of the most significant benefits from the involvement of the public in the assessment process is that in previous efforts of this kind, an ongoing citizen participation in environmental planning was established through the involvement of citizens in the assessment of a project. The involvement of the general public in total environmental planning has been one of the most important goals of the body of environmental legislation passed in the United States in recent years.

5-3.3   Depth of Assessment

In developing an overall approach to environmental assessment, significant attention must be given to the total importance of the project and its probable effects on the environment under investigation. It is necessary to determine, at the outset, whether or not the proposed activity is likely to have significant secondary or indirect effects on the environment. These secondary or indirect effects must be considered as well as the primary or direct effects. For example, the construction and operation of a nuclear power plant would have a very significant effect on the environment in the area where the uranium ores were mined and processed to create the fuel to use in the plant. These mining and processing activities could be considered separate from the construction and operation of the power plant itself, since they would take place at a remote location, but, according to the requirements of the National Environmental Policy Act, these secondary effects must be evaluated. Such indirect effects must be evaluated, in any case, for the environmental assessment to be complete and comprehensive.

Some distinction should be made between short-term and long-term effects. Short-term effects are associated most frequently with construction activities. Areas of impact are related to surface water control, groundwater control, erosion, slope stability, etc. On the other hand, long-term effects associated with the operation and maintenance of a particular facility generally cause impact on the total area surrounding the proposed facility. For example, construction and operation of a fossil fuel power plant could lead to changes in the temperature and chemical characteristics of surface waters and groundwaters in the region surrounding the plant. These changes would constitute long-term alterations in the environment. To carry the example of a power plant further, long-term operations of a plant could lead to the disposal of waste materials from fuel combustion and from air pollution control. The disposal of ash or air pollution control sludges on land could render the land surface unavailable for future use and could lead to the migration of leachates into groundwater supplies (Refs. 5-16, 5-17, 5-36). These long-term effects obviously should be included in any environmental assessments. Processing of fuel for power plants also could produce significant long-term effects. These effects could include water consumption in coal gasification plants through the removal of water from the region surrounding the gasification plant, and release of that water into the atmosphere during combustion where the consumer used the gas at the far end of the pipeline. The effects of such resource depletion in remote areas should be included in any comprehensive environmental assessment. Also, with respect to power plants, the particular mode of power transmission selected for a given plant must be evaluated for environmental effects. Construction and operation of transmission lines for distributing electrical power throughout a region may have serious effects on the local environment (Ref. 5-5). The example of a power plant used in this section can be altered to show the necessities for similar consideration of indirect and long-term effects for many other classes of activity.

5-3.4   Summary on Methods

The methodology for impact assessment has been described in the preceding sections. A method has been recommended: preparation of a matrix, development of a checklist, selection of critical parameters, and involvement of the public. Detailed procedures to attain these general objectives have been described. These techniques and methods are the tools which must be used in the assessment operation. The tools are of little use if competent personnel are not found to engage in assessment investigations and studies. The organization and management of assessment teams is a very complex activity. This important activity is analyzed and described in the following sections of this chapter.

5-4. PERSONNEL ORGANIZATION

In describing methodology to accomplish assessment of environmental impact of proposed facilities or operations, a necessary development is the description of personnel organization. After questions of "What should be in an environmental assessment?", and "How do you write an environmental assessment?", the next most important question is "Who should write the environmental assessment?" The easiest answer to the last question is a question itself: "Who is qualified to write such an assessment?" It should be obvious, from the foregoing description of the activities necessary for adequately assessing the environmental effects of a proposed action, that no single individual would be fully qualified to prepare a comprehensive environmental assessment. A team effort, therefore, is needed. The language of the CEQ guidelines for the preparation of environmental impact statements specifically calls for an interdisciplinary team approach to the preparation of impact assessments. In addition to being required by law, a team effort, with emphasis on integration of effort, is the most practicable and efficient means of accomplishing the complex task of assessment. The types of individuals who must comprise the assessment team can be determined, at least as far as discipline or profession is concerned, by examining the categories of investigation mentioned previously in connection with assessment methodologies. Additionally, the procedure through which assessments are prepared and reviewed can indicate the necessity for input from additional professions and disciplines. From the inception of preliminary planning operations for a given facility through the development of a complete environmental impact assessment, a large number of people should be involved in the preparation and review of the assessments and plans.

5-4.1 Personnel for Preliminary Operations

During the preliminary planning phases for a particular project, various groups of individuals will be involved. The project initiators, be they private individuals or agency personnel, arrive at a concept for a particular project. At this stage in the development of

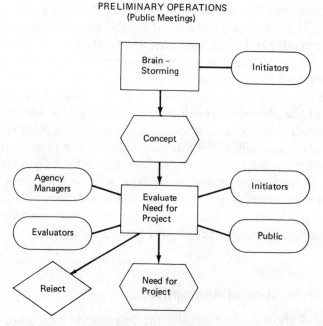

Fig. 5-3 Although project initiators are responsible for developing the basic concept of a given project, evaluation of the need for such a project must be accomplished by members of several different groups, including the public.

any project, as well as later in assessment preparations and review, the opinion of the public should be solicited. A public meeting should be held and the project initiators should be there to present their case. In some instances, a series of meetings with members of the public will be required. The purpose of the public meetings is shown conceptually in Figure 5-3. After the project initiators have developed a concept, the purpose of the public meetings is to evaluate the need for, or the lack of need for, the project as proposed. In this stage of activity, personnel from any federal agencies involved should take part, and these personnel should occupy responsible managerial positions within the appropriate agencies. At this time, the project initiators should present their justifications for the proposed project, for review by the public. Additionally, at this stage, professional evaluators should be included in the meetings and in conferences and planning sessions associated with such public meetings. If the dialogue established in the public meetings indicates that there is a valid need for the proposed project, or if it is determined that a need may exist and further investigation is warranted, the next stage in assessment can be undertaken. If it is shown that no clear and serious need exists for the proposed project, the planning and assessment activity should cease with a rejection of the proposed development. The result of the preliminary operations generally is characterized by a decision that the project is necessary and will fulfill a requirement of the inhabitants of the region wherein the project is to be located, or by a decision that the project is unnecessary and should be rejected. If the project is deemed to be necessary, the various modes of operation and development which can be undertaken to construct and operate the facility or complete the project should be evaluated.

5-4.2 Evaluation of Alternatives

Figure 5-4 shows schematically the operations which are included in the evaluation and identification of alternative modes of action for accomplishment of a proposed project. Project alternatives can consist of different modes of operation, various different locations for essentially identical operations, various different initiation times and periods of duration for the operations entailed in the project, etc. One of the alternatives which must be considered at this point, as mentioned previously, is the maintenance of the status quo. In other words, the alternative of carrying the project development no further should be included in the evaluation at this stage. During this initial evaluation process, the project initiators should take part in order to identify and describe alternative modes of operation which have been developed in the preliminary concept they have formulated. Also in this review and evaluation procedure, the public should be involved and citizen opinions should be solicited. A special effort should be made to contact special-interest groups of concerned citizens in order to assess their input concerning the possible modes of operations and the various locations and durations for a particular project. Additionally, it is obvious that managerial personnel from a federal agency which regu-

IDENTIFICATION OF ALTERNATIVES

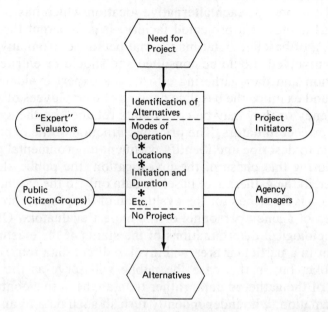

Fig. 5-4   Identification of alternatives requires the cooperation of citizen groups and federal agency personnel (where applicable) in addition to project initiators and assessors so that non-feasible and controversial alternatives can be identified early.

lates activity in a particular area of the environment affected by the project should also be included in this stage of the evaluation, if any federal jurisdictions are involved in the proposed project planning. Expert evaluators also may function well in this stage of the operation in evaluating the various modes of operation, locations, etc. Additionally, the expert evaluators may propose alternatives which have not been considered previously by the project initiators, the agency managers, or the public. The involvement of all these different persons in the identification of alternatives is shown in Figure 5-4. After alternatives to the proposed project have been identified, each alternative should be subjected to a detailed examination. In this examination, the principal group of personnel involved will be the expert evaluators mentioned above. For each alternative under investigation, the alternative project setting must be fully described and evaluated without that particular alternative included. In other words, baseline data must be gathered concerning the environment at each alternative location which has been proposed as a site for a projected facility. It is apparent that in this gathering of baseline data, managerial personnel from any federal agency involved should be consulted and should be enlisted in the evaluation and data-gathering effort. The expert evaluators who gather and examine the baseline data may be employees of a particular agency or they may be consultants retained by the agency or by the project initiators. The principal function of the expert evaluators is to describe and identify pertinent environmental parameters. During this phase of the investigation, the public should be involved in the gathering of baseline data only to the extent that information is furnished for the evaluation of the project through the auspices of agency personnel or the expert evaluators. Of necessity, sociological determinations of the status of the existing environment in a particular area will involve direct data retrieval from the public, but in this case the public will serve as the passive source of the gathered data, rather than as the active contributors of information, who independently furnish such data to an evaluation team.

After the existing conditions have been identified for each alternative scheme, the operation of each alternative project should be identified and described in great detail. During this phase of the investigation, the project or process must be examined with a view toward quantifying the changes in the environment which would be caused if the project is implemented. The project initiators necessarily must be included in this identification process since they possess considerable expertise concerning the various facets of operation of the proposed facility or project. The expert evaluators also necessarily are involved at this stage in the investigation in identifying and expressing impacts in a quantitative fashion, if possible. The public, in general, will not possess significant expertise in either describing the project or in quantifying the changes which will be caused by the proposed alternatives. Thus, in all probability, the lay citizen will not play a significant part in the assessment of alternative actions. However, the expert evaluators should maintain contact with the public in several different capacities in evaluation. The public must be viewed as a comprehensive source of information on existing conditions and on possible effects. However, the investigators must be extremely adept at obtaining this information in such a fashion that it is not compromised by opinion and prejudice. At this stage of the investigation, close coordination should be maintained between the project evaluators and the project initiators. It is possible at this stage in the assessment process to recommend changes in design on the basis of identified impacts. This is one of the most desirable facets of the entire environmental assessment procedure and certainly should not be neglected.

After the various alternatives have been fully identified and described, and quantitative estimates of environmental changes have been made, it is necessary to develop a comprehensive statement of results from all of the foregoing activities. This synthesis of results is indicated by the symbol Σ in Figure 5-5. The result of this synthesis is the development of an assessment report, or, in the case of an environmental impact statement, the development of a draft EIS. The synthesis of assessment results is a very complex and difficult procedure, which requires a great deal of coordination of effort on the part of many individuals.

5-4.3 Coordination of Effort

The various interactions which take place among the personnel involved in the assessment of a particular project are shown in Figure 5-6. These individuals have been active in describing the environmental setting without the various alternatives considered, and in synthesizing the baseline data with the projection of impact. It should be obvious in an examination of Figure 5-6 that a comprehensive team of people will be required for this synthesis effort. A central project evaluator or coordinator should be placed in charge of the preparation of the assessment report. Working with this project manager are various experts who have gathered baseline data, experts in the particular process or project proposed, systems analysts, the project initiators, and other persons (particularly from the public section) who furnish information concerning the possible effects of the proposed alternative projects on all facets of the environment. Also, in this effort, personnel from any federal agency involved in the proposed project should be involved; a close coordi-

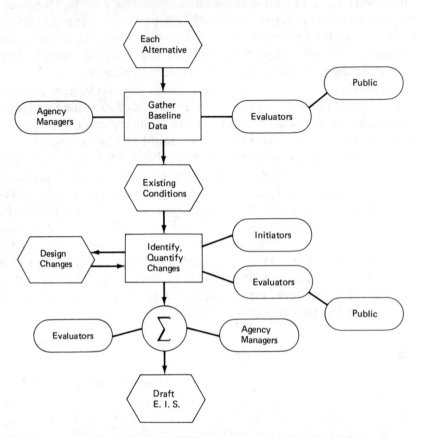

ASSESSMENT OF EACH ALTERNATIVE

Fig. 5-5 Different groups are involved to varying degrees in gathering baseline data, assessing impacts, and synthesizing data (Σ).

nation should be maintained between managerial level personnel from the appropriate federal agencies and the project evaluator. There are several reasons for such close coordination. Close coordination with the federal agency which has jurisdiction in a particular project will give the project evaluator sufficient authority to demand data from project initiators and other concerned personnel. Also, this close contact allows the project evaluator to suggest changes in the design of the proposed facility or project. Since

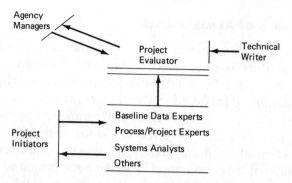

Fig. 5-6 In the synthesis of information for assessment reports, many individuals are involved, but the project evaluator-coordinator bears ultimate responsibility.

these changes can be suggested to persons in authority in the federal agencies which have jurisdiction over the project under investigation, the likelihood of implementation of such changes will be high.

It is virtually mandatory at this stage of the evaluation to have participation by experts in the communication arts, such as technical writers who can assist the project evaluator in developing clear and concise presentations of the results of the assessment process. It is very difficult to express in concise, lucid prose the ideas and assessments formulated by a diverse team of experts. To assist the project evaluator and his communication staff in developing an accurate statement of assessment results, frequent coordination meetings should be held, involving project initiators, evaluators, and agency personnel, as well as members of the public. It is important in this phase of the operation to utilize consensus techniques in discussion among all of the various personnel just mentioned. Consensus techniques yield a much more reliable overall assessment and evaluation than would a combination of the individual evaluations, formulated by the project coordinator. Additionally, if emphasis is placed upon a free communication of ideas and the achievement of consensus concerning the importance and significance of data and projected impacts, the most efficient and comprehensive assessment of environmental conditions and environmental impacts will be attained.

5-4.4   Review of Assessments

In the case of environmental impact statements, a comprehensive review procedure is prescribed by law. Additionally, regulations and guidelines have been promulgated to indicate which federal agencies should be involved in the review process. A comprehensive review may not be required of environmental assessments for projects which do not involve federal funds or do not come under the jurisdiction of a federal agency. However, some state laws require review of environmental assessments prepared under the restrictions and requirements of state environmental policy laws. In the review process, in any case, the comments of the public and other individual experts should be solicited. Too often, draft environmental impact statements are reviewed only by members of federal agency review teams and national environmental group members. The citizens who will be most intimately affected by project activities often never see the environmental impact assessments or the environmental impact statement.

In the review process, various comments on the draft statement or assessment will be generated. After these comments have been received by the project initiators and the federal agencies involved, an effort must be made to reassess the project in light of the comments which have been received. In this reassessment effort, the expert evaluators obviously would be involved. Additionally, the project initiators and federal agency managers, where appropriate, also should take an active part. Not so obviously, the public should be involved in the reassessment of the project under investigation. Members of the public should be included in working committees which are formed to evaluate comments which have been received on impact assessment reports, in order to obtain input from the public in any revision of the proposed project. Through the reassessment procedure, design changes or changes in mode of operation may be recommended to lessen detrimental impacts which were identified or evaluated in the comments made concerning draft environmental assessments or environmental impact statements. These changes can be incorporated into the final assessment or statement by the expert evaluators and the project initiators. It should be remembered that if, during review of the preliminary impact assessment or impact statement, the detrimental aspects of

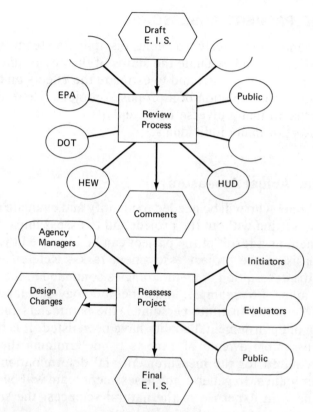

Fig. 5-7 In the review and revision of assessments, public input must be solicited in addition to that of agencies and/or regulatory bodies.

the project are found to be much more significant than initially estimated, it may be appropriate to reject the entire project as conceived. In some instances, total rejection may not be required, but revision of plans for the project may be necessary. In these actions, all of the individuals and groups shown in Figure 5-7 should be involved.

In the foregoing paragraphs, the organization of personnel for the conduct of an environmental assessment has been described in a general way to show the interactions among various groups such as the public, the project initiators, and the expert evaluators. At this point it is pertinent to devote some attention to the organization of the expert evaluation team.

5-5. THE PROJECT ASSESSORS

As mentioned previously, a multi-disciplinary team must be formed to assess and evaluate the status of the environmental setting for a particular project and to estimate the effects on the environment of the completion of a proposed project. There is a need for expertise in many diverse fields and also a need for coordination of effort of many individuals.

5-5.1 The Actual Assessors

The personnel who will be needed to identify and evaluate the environmental setting without the project and to assess impacts on the environment as a result of the project can be identified themselves on the basis of the project assessment tasks. Reference can be made to the description of methodologies contained in the first part of this chapter. For example, in the methodology developed by the Battelle Memorial Institute personnel, the parameters for characterization of environmental quality have been listed. It is relatively easy to assemble a team of experts by determining the type of person required for the measurement and determination of such parameter values. In general, the assessment team will be made up of persons with expertise in the natural sciences, the social sciences, the fine arts, and the applied arts.

With respect to the natural sciences, the physical sciences should be given top priority, in addition to the life sciences. In many of the first impact statements prepared after the passage of the National Environmental Policy Act of 1969, extreme emphasis was placed upon evaluation of the environmental setting by life scientists, including biologists, zoologists, etc. In a comprehensive assessment, the physical sciences should be utilized in evaluating the status of the environmental setting. Team members should have expertise in geography, geology, geophysics, pedology (soil science), and many other relevant physical sciences. It is highly desirable, if possible, to employ persons in the assessment effort who have expertise in more than one of the physical or life sciences, since a close coordination can thus be established between sectors of the assessment team.

The cultural effects of a particular activity or proposed project must be assessed as well as the natural physical environmental effects, so social scientists obviously must be included in the environmental assessment team. The sociological impact of proposed facilities has been a neglected area in many of the environmental assessments prepared within the first five years after the passage of NEPA. In this context, the social sciences should be broadly interpreted to include persons with expertise in economics, sociology, psychology, demography, history, anthropology, archaeology, etc. In prior efforts, assessment of the status of the environment without the project and the evaluation of impacts in the socio-economic sphere has been neglected in comparison to the efforts put forward to identify the quality of the physical environment and to evaluate the impacts on the physical environment of proposed actions. One reason for this neglect is the difficulty in expressing socio-economic characteristics in quantitative terms. Thus, a description of the existing environmental setting in terms of socio-economic characteristics is much more difficult than is a description of the physical characteristics of a particular setting. Likewise, expression, in quantitative terms, of impacts in the socio-economic sphere is much more difficult than quantitative estimation of impacts on the physical environment. However, the overall purpose of assessment is to develop a higher quality of life and to benefit society, so the sociological segment of the environmental assessment effort cannot be neglected.

In addition to natural and social scientists, the assessment team should include experts in the fine arts. In this context, various persons should be employed, according to the needs of the particular project, to evaluate the aesthetic quality of an environmental setting and to evaluate the aesthetic impacts of a proposed project. Artists, landscape architects, and other persons with expertise in the fine arts should be involved in such operations.

Last, but not least, experts in the applied arts of planning, architecture, and engineering should be involved to a great extent in any environmental assessment operation. For example, engineers will possess the technical expertise to properly evaluate the various operational modes and configurations for alternative facilities and actions. The complete identification and description of available alternative modes of operation or alternative locations for given operations is a very important part of the overall assessment procedure. Thus, engineers must be deeply involved in the assessment effort. Additionally, the engineers serve as excellent intermediaries between the scientists involved in the assessment and the project initiators, the agency personnel, and members of the public, since engineers, by training, are skilled in the application of the physical and social sciences. Much the same comments can be made concerning other design professionals such as architects and planners. It is obvious that planners should be involved very extensively in the assessment effort, since the project under consideration consists, in essence, of a plan on paper and no physical facilities, at the time of the assessment. The development of a comprehensive plan for the proposed facility and the evaluation of proposed plans must rest with experts in the planning field.

Thus, personnel with expertise in the natural sciences, in the social sciences, in the fine arts, and in the design professions all must be involved in the assessment operation. The scientists and experts in the fine arts can identify and quantify the characteristics of the existing environment and the probable impacts of proposed actions, while the design professionals, in addition to identifying effects of proposed actions, can also identify other modes of operation and feasible alternative locations for proposed activities. Design professionals can also suggest changes in the proposed action as a result of impact identification by the scientists or the experts in the fine arts.

5-5.2 "Secondary" Assessors

All of the personnel listed in the previous section are employed in identification of the immediate effects of proposed actions and in the quantitative expression of the existing environmental settings. However, in any comprehensive assessment, it is necessary to evaluate second-order and third-order effects of proposed actions, and any suggested modifications or changes in design. In order to accomplish this evaluation of secondary and tertiary effects, it is desirable to assemble another group of evaluators to undertake this task. In essence, the group of assessors investigating the primary effects of the proposed action will be preoccupied with quantitatively identifying and describing the existing environmental setting and the effects, on a first-order or primary basis, of the proposed action. If a separate group is formed to identify and evaluate second-order effects, an independent viewpoint can be achieved regarding the importance of such secondary effects. If one group is assigned the task of evaluating primary, secondary, and tertiary effects, there is a great tendency to devote the major part of the effort towards evaluation of the primary effects, and little emphasis is placed on secondary or tertiary impacts. The most effective way to prevent such neglect of a necessary operation is through the formation of a secondary group which, although working closely with the assessors of primary impacts, should form entirely independent opinions and assessments of secondary and tertiary effects. A very significant effort will be required to insure close coordination between these two groups, but any disadvantages in communication problems will be more than offset through the achievement of adequate evaluation of secondary and tertiary effects.

5-6. ASSESSMENT MANAGERS

Because of the wide variety of backgrounds and areas of expertise included in the personnel of an assessment team, a comprehensive management effort must be exerted in order to attain an assessment in an efficient and effective manner. Without adequate coordination among the various groups assessing the state of the environment and the impact of a proposed facility on various segments of the environment in various degrees (primary vs. secondary vs. tertiary), duplication of effort can occur and at the same time some areas of investigation may be neglected. The task of the management staff in performing an environmental assessment is essentially the application of managerial techniques and principles of systems science to the coordination of effort and to the communication of information among members of the multidisciplinary team.

5-6.1 The Project Assessment Director

The most crucial factor in the management effort is the project coordinator or director. This individual must be a skilled manager who is familiar with all of the areas of investigation included in the assessment. It is not necessary for the manager to be an expert in all of these areas, but he must be sufficiently knowledgeable concerning a wide range of topics that he can converse intelligently and comprehensibly with all members of the assessment team. Additionally, the project assessment director must be completely familiar with all facets of the proposed action and the contemplated project. He must be aware of all of the characteristics of the alternatives which have been proposed in lieu of the contemplated action. In order to become familiar with all aspects of various alternatives, the director must gain the confidence of the project initiators and designers. It is necessary for the director/coordinator to obtain detailed and comprehensive information on all facets of the project. Thus, he must be on good terms with the project initiators and the project owners. Additionally, the project director must seek out personnel in any federal agencies which may have jurisdiction over a proposed activity or who may review an assessment report on a proposed project. It is necessary for the director to obtain information from agency personnel on agency policies and regulations which limit or restrict the modes of operation or the location of proposed facilities. Finally, the project director must seek contact with the public, and especially with intervenor groups who seek to represent the public and protect citizen interests in environmental matters. The project director must be aware of controversy and must be knowledgeable concerning arguments of all sides in such controversies. Ideally, the project manager or director will be a skilled individual who can communicate with potentially hostile groups, bring them together in cooperative meetings, and achieve consensus of opinion on controversial issues. While this is the ideal and should not be expected from real human beings, it is desirable for the project director to have such capabilities as his goals. Since it is almost impossible for one man to have all of these capabilities, the director must assemble a team of management and communication experts to assist him in his efforts.

5-6.2 Communications Within the Assessment Team

Although it may appear to be logical to assume that close communication would exist among the expert evaluators who prepare environmental assessments, such is not always the case. There is a serious need for exchange of factual information and an equally serious requirement for exchange of ideas and concepts concerning environmental quality and social priorities. The members of the assessment team must develop mutual respect for the capabilities of each team member and for the rights of each team member to have independent opinions and views. There is no room on an assessment team for an individual who wishes to climb into an "ivory tower" and prepare an isolated segment of an overall report. Every member of the team must be flexible and willing to accept inputs wherever he may find them. In some cases the inputs will find him. In establishing communications within the team, the project director can be invaluable. Additionally, on very large projects where the assessment work is a heavily budgeted item, professional expertise may be sought in furthering communications within large assessment teams made up of individuals from various disciplines and even from various institutions or consulting offices. The necessity for comprehensive exchange of information and ideas within the team at all stages of the assessment process cannot be overemphasized.

5-6.3 Communication with Project Initiators

One of the activities which is most profitable but most often neglected in assessment operations is the establishment of communications with the project initiators. A direct link should be established by the project director, through his own office, between the project initiators and the members of the evaluation and assessment team. It is absolutely necessary for the assessment team members to obtain all available information on the proposed action and all alternative actions. On the other hand, it is extremely important for the project initiator to be made aware of possible impacts and detrimental effects of features in the proposed design or facility.

If the initiator is made aware of such undesirable effects, he may alter and modify the design or the operation mode for a project to eliminate such detrimental effects. If modification is possible, full and accurate information on such modifications must be furnished by the initiators to the assessment team members. The process of making modifications in response to identification of detrimental impacts is "a two-way street."

All too often, project initiators view the assessment team members with suspicion, and in some cases with outright hostility. The project initiators feel a proprietary pride in their concepts and designs, and quickly rise to defend their proposed courses of action. This hostility is a very detrimental feature since it disrupts communication between initiators and evaluators. Additionally, such feelings of hostility tend to preclude modifications in design in response to identified detrimental impacts. A designer who is deeply enmeshed in defending his design will be reluctant to alter that design in response to comments from the evaluators whom he regards as potential enemies.

Likewise, all too often, members of the assessment team view their role as saviors of the environment who will defend endangered species, including man, against the evil machinations of the project initiators. It is mandatory for the evaluators to accept a design as the best effort from the project initiators. If features of the design are found to produce detrimental effects, such detrimental effects should be enumerated, identified, and described in an objective fashion. The design should be criticized, and not the designer. Much benefit can be gained from activity on the part of the project director and any auxiliary personnel to whom he turns for aid in breaking down emotional and psychological barriers between project initiators and environmental assessors. The project director must always remember that he is dealing with human beings on both sides of the issue, and he must take full cognizance of the frailities of human nature in his attempts to bring about cooperation and communication between various groups involved in the project and in the assessment effort.

5-6.4 Communications With Others

The primary purpose of environmental assessment is to improve the quality of life by allowing the public to make educated choices among various alternative courses of action. Therefore, it is absolutely necessary for the project manager and the project assessors to communicate their findings concerning the state of the environment and probable impacts of proposed actions to regulatory agencies, to various environmentalist groups, and to the public in general. Regulatory agencies may have jurisdiction over certain features of a proposed project and may act to limit implementation of a project on the basis of identified detrimental impacts. On the other hand, regulatory agencies may limit implementation of a project when they do not have comprehensive information concerning impacts. The philosophy taken by agency personnel in such a case is that if some uncertainty exists as to possible detrimental effects of a contemplated action, it is better to be safe; i.e., to delay or prevent the action until all effects of the proposed activity can be identified. Therefore, it is mandatory for the assessment personnel to disclose their findings at the earliest possible time to regulatory agency officials. Likewise, it is necessary for assessment team members to question officials of regulatory agencies in order to determine agency policies as they pertain to the project under investigation. These regulatory policies can be extremely limiting to the implementation of projected operations and facilities.

In addition to obtaining input from agency personnel, it is necessary to communicate with the public. The necessity to communicate with the public is apparent in the need for public input in environmental planning. Involvement of the public in such planning is part and parcel of the overall philosophy of the National Environmental Policy Act of 1969 and other federal environmental legislation. A practical advantage is gained through early communication with the public concerning projected activities since, in many cases, members of the public react hostilely toward projects of which they are ignorant. In these cases, failure on the part of project initiators or assessment directors to convey to the public adequate information concerning the features of the project and the possible effects of the project create a natural reaction of fear and hostility in the minds of the public. It is in the nature of man to fear the unknown and, from a practical viewpoint, it is necessary for a citizen to resist and oppose any environmental action which could have a detrimental effect and about which he knows little. The involvement of citizens in project planning and project assessment at an early stage can do much to alleviate such hostility and can also make initiators and evaluators aware of public concerns with respect to the proposed action. In many cases, public concerns can point toward controversial features of proposed activities, and in some cases public fears of detrimental consequences from particular actions have proved to be well-grounded. A useful concept in understanding the involvement of the public in environmental assessment activities is the "fishbowl" concept. In essence, the project planning and assessment activities should be arranged so that they take place in a communications "fishbowl," open to the view of all concerned citizens.

It may prove difficult to elicit citizen interest and citizen response to proposed activities in early planning states. For this reason, many project initiators, through bitter experience, tend to be very skeptical concerning public participation in environmental planning and assessment. Nevertheless, every effort must be made to obtain input from the citizen. In previous failures to obtain citizen input, the fault may have lain with the communications experts whose duties it was to obtain citizen response and to make the public aware of planned activities. In many cases, no such experts were ever retained by project initiators, and communication efforts with the public were ill-conceived at best.

It is also necessary to establish contact, at an early stage, with environmentalist groups which have been organized to preserve environmental quality and to protect citizen rights to a clean environment. Organized national groups dedicated to environmental preservation and conservation often are the most vocal and dedicated opponents of proposed actions. In many cases, such national groups have even earned the enmity of local residents near a proposed project, through their well-publicized campaigns against projects which the groups have evaluated as detrimental to the environment. There is a feeling at times among local residents that such national groups are merely "sticking their noses into our business." On the other hand, national environmental groups are likely to be extremely well-informed concerning the advantages and disadvantages of various modes of utilization of natural resources. Additionally, it is quite common for environmentalist groups to employ expert consultants and advisors to evaluate particular actions and to guide their activities in either supporting or protesting against a contemplated action. Because such groups often are well informed, almost always are vocal, and can command significant publicity, it behooves the assessment director to establish contact with such groups and to learn their concerns at an early date in the assessment process. Environmentalist groups have been extremely successful in stopping implementation of projects through court actions, and the project evaluator would be well advised to learn the points of concern entertained by the environmentalist groups interested in a particular project. Even if it is not possible to allay the fears of conservationists or to modify designs to eliminate undesirable features, early communication with environmental groups will pinpoint controversial issues in projected actions and will inform the assessment team of possible areas of concern and difficulty.

5-7. AUXILIARY PERSONNEL

In addition to the members of the assessment team previously mentioned, and in addition to the managers of the impact assessment effort, certain auxiliary personnel may be required in assessment operations, especially where a proposed action involves a considerable commitment of natural resources and financial resources, or is likely to have extremely far-reaching effects. On almost any assessment effort, certain auxiliary personnel are required. These individuals will be a part of the assessment team and their principal functions will be to provide coordination and communication, both within the team group and with external personnel. Additionally, certain general support personnel will be required.

5-7.1 In-House Coordinators

In addition to the project manager or director mentioned in the preceding section, it may be highly desirable (especially for large assessment efforts) to employ specialists in communications and in behavioral analysis/modification to facilitate cooperation and coordination of the efforts of the members of the study team. Some of the individuals employed in this effort will be clerical or secretarial workers whose function is to distribute and disseminate data throughout the study team. This function may seem to be rather mundane, but the maintenance of complete records and the simple distribution of collected data to all members of the assessment is a formidable task. The task of keeping all members of the assessment team informed concerning the progress of the assessment effort, possible team meetings, and public meetings is very difficult and very important. Thus, a significant need exists for clerical managers who facilitate the flow of information among members of the assessment team.

In addition to the clerical workers who are in charge of the scheduling and information distribution, in extensive assessment efforts it may be highly desirable to employ specialists in group counseling who can work with all of the members of the assessment team to encourage and improve their cooperative efforts. The time and effort spent in developing group cooperation and coordination through the conduct of in-house seminars and similar endeavors generally pays great dividends in the form of more efficient and comprehensive evaluations of existing conditions and possible impacts. Certainly, a significant effort toward improving group dynamics in the assessment team should be considered in large assessment projects.

5-7.2 Public Relations Experts

In addition to disseminating information among the members of the assessment team, it is mandatory in comprehensive assessment operations to organize public meetings and educational programs to inform the public concerning projected activities and proposed actions. Additionally, it is important to inform the public concerning the activities of the assessment team. After the assessment is in progress, it is important to continue to have public meetings to inform the public about identified effects of the proposed actions and alternative actions, including both detrimental and beneficial effects. In all of these efforts, it is necessary to inform the public and, if possible, to educate the public. A fine distinction can be drawn between these two activities. The public may be informed through the dissemination of notices and through factual statements at public meetings and similar activities. The public can be considered to be educated only when it is obvious that individual citizens have a comprehensive understanding of the issues involved in a particular assessment. In order to determine the success of the communications efforts directed at the public, it is necessary to monitor public awareness through questionnaires and individual interviews with persons in the areas affected by the proposed activities. At times, it may seem that the majority of the public is apathetic or hostile toward a contemplated action, and it is easy for the assessment team members to become complacent or cynical concerning public input. Such attitudes are self-defeating, since the entire purpose of the assessment effort is to inform the public so that they may make intelligent choices among alternatives. It is mandatory to do everything possible to communicate with the public concerning the issues involved in a particular project and to elicit public reaction to planned activities and to the assessment of probable impacts.

In addition to communication with the public at large, it is necessary to establish liaison with agencies and environmentalist groups who may have jurisdiction over a particular project or who may have interest in that project. For this purpose, and for communication with the public, individuals specifically trained in developing communication skills should be retained. The importance of good communications with the public and with interested groups as well as regulatory agencies should not be underestimated. It is a wise investment to retain a professional public relations expert to plan and coordinate such communications activities.

5-7.3 Writers and Illustrators

The end product of the assessment operation in almost every case is a written document. As such, it is subject to all of the limitations of any other form of written communication. In order for the assessment effort to be worthwhile, the report of the assessment or the environmental impact statement must be readable. It must be readily comprehensible to the lay public. In order to facilitate communication via a written document, it is extremely beneficial to employ professional writers and illustrators to prepare the final assessment report or the environmental impact statement. These writers and illustrators, however, must not be sequestered in a remote office and given voluminous study reports to summarize and distill into a readable and comprehensive assessment report or impact statement. It is necessary that writers and illustrators form close ties with the members of the assessment team in order that they appreciate the relative importance of various parameters and the significance of the results of the assessment effort. If a proper combination is achieved between the gathering and reporting of information by the assessment staff and the distillation and collation of such information by professional writers and illustrators, it is possible to produce concise and lucid assessment reports and impact statements. One of the most grievous failings of the impact statements which have been prepared since the passage of the National Environmental Policy Act of 1969 has been the fact that most of the statements are verbose and so long that they discourage all but the most persistent reader. If an assessment report or impact statement is not read by the public, then the very purpose of the assessment operation is defeated.

5-7.4 General Support Personnel

In addition to all the individuals named in previous sections, in any comprehensive assessment operation, a large number of support personnel will be required. These personnel will include laboratory technicians, field technicians, surveyors, draftsmen, clerical workers, auditors, etc. These persons form the support cadre which customarily is found in all scientific or engineering surveys. However, in the conduct of environmental assessments, to date, many assessment efforts have been carried out with meager and insufficient support staffs. The lack of adequate support from field personnel and laboratory technicians often leads to the gathering of insufficient amounts of information or to the reporting of unreliable results. Insufficient office staff in clerical posts may significantly delay the completion and distribution of an assessment report or an impact statement. After this problem has been mentioned, it may appear to be somewhat simplistic and the solution is obvious. Simply develop an adequate support staff! However, even though this matter may appear to be obvious as it is stated in this text, it has been the authors' experience that many assessment endeavors are attempted with insufficient support staff.

5-8. SUMMARY ON PERSONNEL

In summary, the personnel required for a comprehensive environmental assessment include members of the assessment team, project managers and directors, and a body of support personnel.

5-8.1. The Assessors

The members of the assessment team can be selected through an examination of the investigations required in the assessment itself. In order to determine the quality of the existing environment, and in order to assess the effects of proposed changes, it will be necessary to employ physical scientists and life scientists to evaluate the condition of the physical environment and the biological elements within that environment. Additionally, specialists in the fine arts will be required in order to assess the aesthetic features of the existing environment and the aesthetic impacts of the proposed activities. To furnish an assessment of socio-economic conditions and to evaluate possible socio-economic impacts, a team of social scientists will be required. Finally, design professionals, including architects, planners, and engineers, will be required in order to fully evaluate and describe the proposed activity and to systematically analyze the results of the investigations carried out by the physical and social scientists.

5-8.2 The Project Director

Ideally, the overall responsibility for the development of the environmental assessment should be delegated to a jack-of-all-trades who, if possible, is master of several of those trades. In other words, the project coordinator must be conversant with the various disciplines represented by members of the assessment team. He may be expert in one or more of the disciplines represented. Indeed, it is hardly likely that any one individual would be expert in more than two separate disciplines. However, the project director must have a working knowledge of the major principles involved in the activities of each of his team members. Additionally, the project manager must be adept at developing cooperation among members of the assessment team. To facilitate such cooperation, the project manager must establish well-defined channels of communication among the team members. It is absolutely necessary that members of the assessment team develop mutual respect; team members must be receptive to the ideas and opinions of their fellow workers. In addition, the project director must form close ties and excellent communications with agency personnel, with project initiators, with concerned environmentalist groups, and with the public at large. To assist in this effort, he should employ professional communication experts.

5-8.3   The Support Group

In addition to the members of the assessment team drawn from physical and social sciences and the fine and applied arts, the project director must be assisted by various experts in specific areas of communication and information retrieval and distribution. It is highly advisable to have in-house coordinators who will maintain adequate records and disseminate information among the members of the study group. Additionally, it is necessary to have a clerical staff who will supervise scheduling of all activities. On extensive assessment studies, it is beneficial to employ experts in group dynamics to improve the efficiency of the assessment study team in functioning as a cohesive group. In addition, public relations experts should be retained to organize educational programs to disseminate information to the public. Such programs can include public meetings and other activities. These public relations experts should monitor public awareness through interviews and questionnaires. The public relations experts involved in assessment work should also establish communication with other groups, such as agency personnel, project initiators, and environmental groups. To produce a readable and manageable final assessment report or impact statement, it is virtually mandatory to employ professional writers and illustrators to distill and clarify the copious information gathered and reported by assessment team members during the course of the assessment study. One of the principal failings of environmental impact statements to date has been their voluminous and intractable character. If a statement is too large or too boring to be read, its basic purpose has been defeated. Finally, in addition to all the support personnel mentioned above, various technicians and clerical staff should be retained to assist the other members of the assessment team. Unfortunately, many assessments have been carried out, to date, with inadequate support staffs; incomplete or unreliable information has been obtained in some cases.

5-9. SUMMARY

In this chapter, the methods for environmental assessment have been described. It was pointed out that many traditional methods of technical and economic evaluation are not suitable for use in environmental assessment because of the difficulty in assigning numerical values to environmental quality parameters. Graphical techniques may be used, particularly as communication tools, in portraying the physical characteristics of a proposed activity and in displaying the degree and location of expected impact. Graphical techniques are particularly useful in public meetings and in similar activities designed to create public interest in environmental planning processes. However, graphical techniques are not sufficiently discriminatory to satisfy all the needs of comprehensive environmental assessments.

Numerical techniques have been developed for environmental studies, but such techniques tend to be either oversimplified, or insufficiently flexible for general use. Some numerical methods have been modified so that they are useful in accomplishing portions of the total environmental assessment task. For example, matrices may be utilized to identify possible areas of impact and to graphically display interactions between contemplated activities and particularly valuable environmental attributes. Checklist systems can be employed to insure that no environmental characteristic is neglected in the completion of an environmental assessment.

Many quantitative systems have been developed to produce a numerical value representative of the total impact of a proposed activity or a suggested alternative. Most of the quantitative systems which have been developed are not sufficiently flexible for application to large numbers of activities in various locations. Additionally, many of the quantitative systems which have been developed give a false impression of the accuracy of the assessment process. Comparison of numbers representing degrees of impact may lead to a false impression concerning the accuracy with which the impact is predicted and may lead to the neglect of certain particular areas where impact is severe, since the numerical representation of the *total* impact may not indicate such severity.

A total approach for environmental assessment has been suggested in this chapter. Use of a matrix for preliminary identification of impact has been recommended. The development of a comprehensive checklist also is a very worthwhile activity. Involvement of the public in the assessment process at the earliest possible stage is a virtual necessity in any comprehensive study. Techniques for insuring significant citizen participation have been outlined. The necessity to include indirect and long-term effects of particular projects in the assessment study has been mentioned.

The second portion of this chapter has been devoted to a description of methods for organizing assessment teams. It is absolutely necessary that the assessment team constitute an interdisciplinary study group rather than a multidisciplinary organization. In other words, communication must be frequent and candid among members of the assessment group, and between members of that group and the project initiators and the general public. If possible, members of the assessment group should possess expertise in a number of disciplines so that truly interdisciplinary investigations can be achieved.

One of the most vital activities in the entire assessment process is the overall management of the evaluation team. The evaluation team manager must possess a working knowledge of each of the fields of expertise represented by members of the assessment team. He must be a skilled leader and a shrewd manager, capable of dealing fairly and cordially with project initiators, impact evaluators, and concerned citizens. On projects of significant importance, the assessment team should include specialists who can contribute their talents in developing graphical displays of critical information, in preparing clear and lucid descriptions of proposed activities and expected impacts, and in developing effective means of communication with the general public. The overall assessment objective will not be obtained without effective management on the part of the assessment team leader or without full cooperation among project initiators, impact evaluators, and interested citizens.

REFERENCES

5-1. Adkins, W. G., *et al.,* "Social, Economic, and Environmental Factors in Highway Decision Making," *148 Progress Report,* Study 2-1-71-148, Texas Highway Department and Fed. Hwy. Adm., Washington, 1971.
5-2. Arnstein, S. R. and Winder, Jr., S. S., "Discussion of Potential Uses of Interactive Computer Graphics in Citizen Participation," *Trans. Res. Record 553,* Trans. Res. Board, Washington, 1975.
5-3. Battelle Memorial Institute, Columbus Laboratories, "Final Report on an 'Environmental Evaluation System for Water Resource Planning' to the Bureau of Reclamation, Department of the Interior," January, 1972.
5-4. Carter, Everett C., Haefner, Lonnie E., and Hall, Jerome W., "An Informational Report on Techniques for Evaluating Factors Relevant to Decision Making on Highway Locations," DOT, Fed. Hwy. Admin., Off. Env. Policy, Washington, November, 1972.
5-5. Caswell, Alfred E., Nazare, Charles V., Berger, Richard J., and Torri, Joseph F., "Environmental-Geotechnical Considerations in the Routing of EHV Transmission Lines," *Proc. of the ASCE Specialty Conf. on Elec. Power and CE,* Boulder, August, 1974.
5-6. Curry, M., "What Role for Citizens?" *Trans. Res. Record 555,* Trans. Res. Board, Washington, 1975.
5-7. Dries, David A., "Environmental Impact Assessment Methodology," M.Eng. Thesis, Speed Scientific School, University of Louisville, May, 1974.
5-8. Engelen, R. E., and Stuart, D. G., "Transport Technologies and Urban Structure: A Framework for Evaluation," paper presented at the ASCE-ASME National Meeting on Transportation Engineering, Seattle, July, 1971.
5-9. Environmental Assessment Group, University of Louisville, "Environmental Impact Assessment of the Cannelton Locks, Dam and Pool," University of Louisville, Louisville, November, 1973.

5-10. _____, "Environmental Impact Assessment of the Newburgh Locks, Dam and Pool," June, 1974.

5-11. _____, "Environmental Impact Assessment of the Uniontown Lock, Dam and Pool," November, 1974.

5-12. _____, "Environmental Impact Assessment Study Report, I-65 Project," January, 1975.

5-13. Feeser, L. J., and Ewald, R. H., "Computer Graphics and Public Hearings," *Trans. Res. Record 553,* Trans. Res. Board, Washington, 1975.

5-14. Girand, J., "Technical Movie of Papago Freeway," *ASCE Journal,* Trans. Eng. Div., Vol. 97, TE4, November, 1971.

5-15. Haefner, L. E., and Redding, M. J., "An Analytical Structure of Community Public Works Decision Process," *Papers and Proceedings of the Joint Conf. Env. Des. Res. Assoc. and Amer. Inst. Arch.,* Irvine, California, January, 1972.

5-16. Hagerty, D. J., and Pavoni, J. L., "Geologic Aspects of Landfill Refuse Disposal," *Engineering Geology,* Vol. 7, March, 1974, pp. 219–229.

5-17. Hagerty, D. J., Pavoni, J. L., and Heer, Jr., J. E., *Solid Waste Management,* Van Nostrand Reinhold, New York, 1973.

5-18. Hagerty, D. J., and Heer, Jr., J. E., "Impact Assessment of Power Plants," *ASCE Journal,* Power Division, Vol. 102, No. 1, January, 1976.

5-19. Harper, B. A., "Achieving Effective Citizen Participation," *Trans. Res. Record 555,* Trans. Res. Board, Washington, 1975.

5-20. Jain, R. K., Urban, L. V., and Stacey, G. S., *Handbook for Environmental Impact Analysis,* Dept. of Army, CERL, Champaign, Illinois, (Draft), April, 1974.

5-21. Leopold, Luna B., Clarke, Frank E., Hanshaw, Bruce B., and Balsley, James R., "A Procedure for Evaluating Environmental Impact," Geological Survey Circular 645, U.S. Govt. Printing Office, Washington, 1971.

5-22. McHarg, Ian L., *Design With Nature,* Natural History Press, Garden City, New York, 1969.

5-23. National Thermal Pollution Research Program Staff, "Reviewing Environmental Impact Statements-Thermal Power Plant Cooling Water Systems," Report, USEPA, Corvallis, Oregon (In Press).

5-24. Newmark, Nathan M., "Design Criteria for Nuclear Reactors Subjected to Earthquake Motions," Paper presented at Int. Atomic Energy Agency Meeting, Tokyo, June, 1967.

5-25. Oglesby, C. H., Bishop, B. and Willeke, C. E., "A Method for Decisions Among Freeway Location Alternatives Based on User and Community Consequences," *Hwy. Res. Record 305,* Highway Res. Board, Washington, 1970.

5-26. Redding, M. J., "The Quality of Residential Environments, Preference for Accessibility to Residential Opportunities," Ph.D. Thesis, Northwestern University Technological Institute, Evanston, June, 1970.

5-27. Schimpeler, C. C. and Grecco, W. L., "Systems Evaluation: An Approach Based on Community Structure and Values," *Hwy. Res. Record 238,* Highway Res. Board, Washington, 1968.

5-28. Schlager, K., "The Rank-Based Expected Value Method of Plan Evaluation," *Hwy. Res. Record 238,* Highway Res. Board, Washington, 1968.

5-29. Sheridan, T. B., "Technology for Citizen Participation in Planning," *Trans. Res. Record 553,* Trans. Res. Board, Washington, 1975.

5-30. Stuart, D. G., "Concepts, Approaches and Problems of Applying Interactive Graphics in Community Participation Programs for Urban Transportation Planning," *Trans. Res. Record 553,* Trans. Res. Board, Washington, 1975.

5-31. United States Atomic Energy Commission, "Environmental Statements-Operations," Title 10, Chap, I, Part ii, U.S. Govt. Printing Office, Washington, July, 1972.

5-32. United States Atomic Energy Commission, "Preparation of Environmental Reports for Nuclear Power Plants," Regulatory Guide 4.2, U.S. Govt. Printing Office, Washington, March, 1973.

5-33. United States Atomic Energy Commission, "Licensing of Production and Utilization Facilities," Title 10, Chap. I, Part 50, 38 *Federal Register,* 31279, November 13, 1973.

5-34. United States Department of Agriculture, "Environmental Impact Statements: Proposed Guidelines," 38 *Federal Register,* 31904, November 19, 1973.

5-35. United States Department of Housing and Urban Development, "Handbook of Departmental Policies, Responsibilities and Procedures for Protection and Enhancement of Environmental Quality," 38 *Federal Register,* 19182, July 18, 1973.

5-36. Weeter, Dennis W., Niece, James E., and Di Gioia, Jr., Anthony M., "Water Quality Management-Power Station Coal Ashes," *Proc. of the ASCE Specialty Conf. Elec. Power and CE,* Boulder, August, 1974.

5-37. Werner, Robert R., "Environmental Impact and the U.S. Army Corps of Engineers," Paper presented at Ecological Soc. of America Symposium, Amer. Assoc. Adv. Sci., Washington, December, 1972.

5-38. Winter, John V., "Power Plant Siting," M.Eng. Thesis, Speed Scientific School, University of Louisville, May, 1974.

5-39. "Environmental Quality as an Objective of Water Resources Planning," Special Printing, Comm. Int. and Ins. Affairs, U.S. Senate, U.S. Govt. Printing Office, Washington, 1972.

5-40. Executive Order 11514, "Protection and Enhancement of Environmental Quality," 35 *Federal Register,* 4247, March 7, 1970.

5-41. *Proceedings, Panel Discussion on the Systematic Interdisciplinary Approach in Highway Planning and Design,* ed. MIT Trans. and Comm. Values Proj., Urban Sys. Lab. with Fed. Hwy. Adm., Cambridge, March, 1973.

6

Case Histories

6-1. INTRODUCTION

In order to illustrate the applications of the methodology and organizational procedures described in the preceding chapters, several case histories will now be presented. These case histories illustrate the applications of some of the laws mentioned in Chapter 2, as well as illustrating the utilization of assessment techniques for projects not covered under the provisions of the National Environmental Policy Act of 1969. Included in the cases described are two project assessment studies for Interstate highway projects. In the first study, a projected highway widening was assessed for environmental impact. This case serves as a good example of all of the investigations required under the provisions of NEPA. The second case described is a highway project, too, but it is presented for a different reason than is Case 1.

The technical operations involved in assessing impact in Case 2 are very similar to those in Case 1; Case 2 is described to present a case history of very comprehensive public involvement in environmental planning and evaluation. Although the technical evaluation operations carried out by the professional assessors working on Case 2 were quite similar to the operations carried out in Case 1, the very comprehensive public relations effort employed in Case 2 deserves separate description. This public relations effort was made necessary by the provisions of NEPA, but the need was made intense by the fact that the project in Case 2 was to be constructed in a very heavily populated area.

In Case 3, a different kind of project is presented. In this case history, the assessment of impact for a water resources project is described. Additionally, the impact is described during the planning stages; i.e., the impact assessment is made on the basis of a developed plan for water resources management and not for the actual project to accomplish that management. Because of this particular characteristic, Case 3 is a useful illustration of the assessment of planning procedures and can serve as a model for similar efforts in other fields of endeavor.

In Case 4, a different situation is described, in that the project under assessment was virtually completed at the time of the assessment. This state of affairs was created by the passage of the National Environmental Policy Act of 1969 at a time when this particular project was approximately 75% complete. Nevertheless, court rulings indicated that an impact assessment would be required for this type of project, and so an impact assessment was prepared. This case can serve as a good example of the assessment of one type of situation, that of a project already underway. Cases 3 and 4 represent the extremes of assessment timing; i.e., during planning stages at the earliest possible time in project development, and at a time when a project is virtually complete.

Cases 5 and 6 are devoted to descriptions of projects in which smaller amounts of money and less intensive efforts were involved, in comparison to the first four cases mentioned above. However, these case histories serve to illustrate the application of assessment methodology to localized problems and to small projects. In one of the cases, a federal agency exercised jurisdiction over the project, but no federal monies were involved in the construction or planning stages. In the other case, no federal money or federal jurisdiction was involved in any way.

In Case 5, the application of assessment methodology to evaluating disposal or recycling methods for a particular waste product is illustrated. This problem was highly localized since a change in disposal methods was required by a local pollution agency. On the other hand, in Case 6, a federal agency exercised jurisdiction over a particular project but the project itself was a privately financed water-related development. This case serves to illustrate in a graphic way the changes in design and improvements in planning which can be obtained through early assessment of environmental impact.

The authors have participated in most of the studies upon which these case histories are based. In three of the cases, they participated as members of a University of Louisville environmental assessment team; in other cases, the assessment effort was carried out by the authors as a separate group with auxiliary personnel employed where necessary.

6-2. CASE ONE—HIGHWAY IMPROVEMENT

The cases in this chapter were chosen to present one or more significant points where the agency or its consultant did an unusually good job in preparing and presenting the data, or, in some cases, did an exceptionally poor job in one or more aspects of the project. Each of these case histories describes an actual project which has been proposed; however, several of the projects (because of public opposition or lack of adequate finances) to date have not been completed.

This first case history is concerned with the improvement of a major highway project in a large urban area in the midwest. While this particular project has not been undertaken, primarily because of public opposition, it is still under active consideration and alternatives are currently being investigated to implement the project in a way that would be more acceptable to the community. The project itself consists of upgrading the traffic-carrying capacity of a major highway segment. The segment currently is incorporated into the Interstate highway system, even though parts of it originally were constructed prior to the implementation of the Interstate Highway Act by Congress. As a consequence of the construction of parts of the highway prior to the enactment of the Interstate Highway Act, the highway as constructed, and as it currently exists, is not developed to Interstate standards, either in the matter of capacity (ability to handle the traffic loading as specified), or acceptability from the standpoint of traffic safety.

The original highway was planned as a belt-line around the major residential areas of this midwestern city; the planning was done in the period prior to the beginning of World War II. The entry of the United States into World War II delayed any construction on the highway until 1949. A major segment of the highway (the entire project, complete, contained approximately 25 miles of belt-line) was constructed between the years 1949 and 1958. This section is approximately 13 miles long and thus consists of slightly more than one-half of the entire belt-line. It is a four-lane divided highway throughout its length; however, some two-thirds of it was not built to Interstate standards. The other third was built to Interstate standards, with the exception of the geometry of acceleration and deceleration ramps and what today would be recognized as adequate interchanges.

Since much of it was built before the Interstate System was conceived, it has many deficiencies that would not be tolerated on newer roads. It contains some of the worst accident sites in the urban community that it serves as a by-pass or belt-line. In the years since the roadway was opened, there have been numerous fatal head-on collisions as a result of vehicles crossing the narrow raised median. This condition led to the installation of a double-beam guardrail throughout most of the roadway length in order to reduce the annual fatalities on the highway.

A comprehensive transportation planning report made in late 1969 for the metropolitan region served by this highway recognized that this belt-line was one of the most critical in the region and recommended very strongly that it be expanded to allow an increase in capacity while at the same time improving the safety aspects. This consultant report stated: "Of all of the highway facilities in the metropolitan region, the one which most affects development is the expressway considered in this case. The rapid rate of development along this facility perpetuates congestion. If increased congestion of the facility inhibits growth, it is certainly not apparent. The great attractions of the facility are its local access and the visibility that is provided for establishments that locate along its right-of-way. After detailed study of alternate routes to the north and the south, it is recommended that the expressway be widened."

With this, then, as a background, the local, state and federal officials assigned a high priority to the implementation of a project which would bring the entire belt-line (but in particular the 13-mile section which was constructed prior to 1960) up to current Interstate standards, with an acceptable level of capacity for projected 1975 traffic density and facilities for expanding to predicted 1990 traffic densities. It was under these circumstances that the environmental assessment presented in this case history was undertaken.

6-2.1 Environmental Setting Without the Project

The metropolitan community which the project addressed in this case history is located in the largest city in a midwestern state on a major navigable river. Three of the state's Interstate routes pass through this urban community and converge on its Central Business District. The highway in question is a circumferential Interstate route, planned as a connection around the Central Business District between these three radial Interstates converging in the downtown area. Thus, the highway serves two major, yet entirely different, purposes. It serves as a connector route between the three Interstate highways, without the necessity of a traveller going to the Central Business District for interchanging from one Interstate highway to another and then leaving by a different radial route. Of course, this serves to eliminate some of the through traffic congestion that might otherwise take place unnecessarily in the Central Business District. The second feature of the existing highway, as previously mentioned, is to serve as a distributor of semi-local traffic in the various business and residential areas that have developed around it, simply because of the attraction of its facility and because of the large number of commercial establishments that have developed along its right-of-way.

When the belt-line was opened to traffic in the late 1950's, it very quickly became congested; at the present time, it is used by more than 80,000 vehicles daily. The portion of the expressway being considered for improvement runs east and west between two currently existing federal (non-Interstate) highway routes. In its 13-mile length, it intersects and connects with all the major arteries that converge on the urban area. It presently has 13 closely spaced interchanges, with two additional partial interchanges, in its 13-mile length. The present right-of-way varies from 120 feet in width for approximately two miles of its length, to 220 feet in width for more than half of its length.

As previously related, the existence of the belt-line, even though inadequate from a traffic and safety point of view, has spurred a marginal urban growth in what is now an all too familiar pattern. Major high density residential developments, shopping malls, and industrial plants, placed to take advantage of the transportation benefits and access provided by the highway, soon fringed the expressway from one end to the other. All these developments, of course, added to its traffic load, and levels of traffic were reached at a much earlier date than had been forecast and have continued to grow with almost every passing year.

Many of the problems that exist on the belt-line today were caused by the official policies in effect during the initial construction of the highway. These policies attempted to get the most highway for the money then available, even if some undesirable features had to be included. For example, in several of the interchanges, in the name of economy, structures were allowed to remain within pieces of cloverleaf interchanges, with access to the structures from the interchange ramps. In other cases, existing residential streets were incorporated into the interchanges to serve as ramps. In at least one case, an existing highway bridge over a series of railroad tracks on a residential street was raised and incorporated, unwidened, into the belt-line. This bridge now sits in the area of the narrowest right-of-way, and commercial and residential structures exist within less than 50 feet of the edge of the outer driving lanes.

As previously mentioned, in the 13-mile length of the highway, there are, at present, some 15 interchanges (13 entire, 2 partial).

This is obviously an excessive number for any type of Interstate or expressway traffic, and causes undue congestion at many points along the expressway. For several hours each day, during the morning and evening rush hours, the traffic operates over many areas of the highway in a stop-and-go condition, often with vehicles backing up for a mile or more.

It is this type of environment which has been recognized by the officials of the metropolitan area, as well as the State Department of Transportation and the citizens of the community, as requiring some type of improvement, from the standpoint of expediting the flow of traffic and increasing safety on the highway.

Other environmental problems existing along the highway, without the implementation of the proposed project, are primarily those of noise and air pollution from the traffic, because of the development of commercial and residential areas along the highway since its construction. In certain areas, where the right-of-way was narrow to begin with, many homes and businesses currently are exposed to high levels of traffic generated noise. In addition to the noise, those people working or living near the highway are exposed, in times of heavy traffic, to high levels of air pollution from the cars, buses, and trucks using the highway.

Also, there are located, almost immediately adjacent to the highway, at least six churches and three other structures including a parochial school, a home for orphaned boys, and a recreational area owned by a religious organization. In the length to be improved, there is currently one historic home listed on the National Register of Historical Sites. Other than these structures, most of the nearby residences or businesses are of the usual type found in an urban area adjacent to a heavy-traffic artery.

6-2.2 Impacts of the Proposed Action

In considering the improvement of the 13-mile stretch of highway, it was necessary to develop various alternate solutions; solutions were proposed which consisted of adding additional lanes in each direction to the existing highway. Consideration was given to adding two lanes in each direction to raise the level of traffic capacity to that which would be adequate for the predicted 1975 traffic. Another alternative was to add as many as four lanes in each direction to the existing two in each direction; this proposal would be adequate for the predicted 1990 level of traffic. In addition to these proposals, consideration was given to the installation of a rapid transit system down the existing median, with the addition of two or three new lanes in each direction; presumably, the rapid transit system would relieve some of the traffic from the highway, reducing the demand to such a level that the additional proposed lanes for the 1990 traffic would not be needed. Another proposed solution was to add a parallel highway some distance farther removed from the Central Business District, which, in essence, would be a second belt-line around the city, intersecting the major radials farther from the center of the city. Of course, as required by the National Environmental Policy Act, the do-nothing or null alternative also was considered.

In each of these alternatives, consideration was given in the planning to the impacts of noise, air pollution, dislocation of businesses and families, as well as to the taking of some lands from the previously mentioned church-affiliated properties. The environmental consultants and assessors also addressed themselves to an assessment of the impacts with respect to safety, economy, function, and sociological impact on the community.

In earlier discussions with some of the citizens living immediately adjacent to the highway, two questions were asked: "Do you think the highway should be improved and widened?" and "What is the most disturbing feature of the highway as it exists today?" The single most repeated complaint expressed by those residents living near the highway was directed against the noise generated by the traffic, especially by the large number of heavy trucks using the highway. In this respect, a major assessment and noise study was undertaken to document the changes in the acoustical environment along the highway for each of the proposed alternatives. Sound pressure level measurements were taken along the highway for the existing level of traffic, and, by means of computer programs developed for the Highway Research Board, these existing levels were extrapolated to a predicted level for the 1980 and 1990 traffic. It was found that, in most areas where residential or commercial development has taken place, even with the widened highway and the additional right-of-way that would be taken by the proposed project, the noise levels would be excessive in much of the property immediately adjacent to the project. For this reason, the so-called preferred plan (which was a widening proposal of three additional lanes in each direction for a total of ten lanes on the highway) was presented with the requirement that it would be necessary to develop either an earth berm along the edge of the right-of-way, with a crest barrier wall, or an extremely high barrier wall without the earth berm, in order to alleviate the increased levels of noise predicted for the increased traffic.

A second impact assessed was that on air quality. An extensive study was made to identify the effects of expanding the width of the highway, and thus increasing vehicular traffic. Sampling was done along the existing right-of-way and, again from computerized models, the levels of air pollution were projected for the 1990 traffic. This study, taking into account the development of air pollution devices on vehicles, as currently required by the federal air pollution control laws, predicted that the actual level of air pollution adjacent to the new widened highway would be less than that which existed at the time of the study in 1974. In spite of the additional traffic generated, the level of air pollution from each vehicle would be reduced to such an extent that the cumulative effect actually would be lower.

Consideration was also given to the changing land use patterns which would be caused by the newly widened highway. Eight hundred thirty-five families would be displaced by the improvements proposed as the preferred alternate (number 1). Of these, 262 were classified as "hard displacements;" that is, householders who, because of low income, old age, or other reasons, might have difficulty finding comparable housing without appreciable financial assistance. A total of 62 businesses were predicted to be displaced. It was estimated that, of these, 30 would relocate near their present location, while the remainder would probably relocate at a considerable distance or not relocate at all. As a compensating factor to these business dislocations, it was predicted that, should the project be implemented, major business developments will take place, particularly along the highway in the vicinity of the major interchanges.

A slight drop in tax revenues was predicted because of the loss of businesses and residential properties along the highway; however, the assessment stated that this would be only a short-term loss in that the new predicted developments along the improved transportation facility would, in fact, be more valuable than those lost, and that an increase in the tax base would be the long-term result.

Most of the alternatives were discussed briefly and dismissed as being not politically, economically or sociologically appropriate. The do-nothing option was dismissed as being more costly in the long run than actually improving the highway. This assessment was reached by assigning a dollar cost to the time lost in traffic congestion and also by the dollar volume of business lost in the adjacent areas due to shoppers' unwillingness to put up with the traffic congestion. The more remote outer belt was dismissed as not being feasible. The high cost of the extensive residential and commercial development in this area would require the outer belt to be such an extensive distance from the existing belt-line that it would not appreciably draw traffic from it. It would not solve the problem of excessive traffic demand. The mass transit option also was thoroughly investigated and dismissed as not feasible at this particular time; it was the opinion of the assessors that sufficient traffic would not be drawn to the mass transit system any time in the near future to adequately relieve the congestion on the existing highway without the addition of lanes in either direction. Other alternatives, such as incentives for car pooling, staggered work hours for local businesses, and restricting the type of traffic that could use the existing highway, were dismissed as not being practical from a political point of view. It was felt that there was no indication that such measures would be adopted by any major segment of the community, nor would such measures receive any public support. Thus, any attempt to solve the problem by one of these administrative options was seen to be doomed because of the lack of public support, and therefore, was not considered to be a viable alternative.

Because this project consisted of a widening of an existing highway through a well-developed area, there was little, if any, wildlife in the area, and certainly no endangered species. The flora and fauna found along the existing highway were of the most common types. Trees and shrubs, for the most part, were those which the highway department had planted since the existing highway was constructed. Therefore, the widening of the existing highway would have negligible impact, either positive or negative, on the wildlife of flora and fauna.

6-2.3   Other Aspects (NEPA Requirements)

As mentioned in earlier chapters, NEPA is quite specific in requiring certain aspects, such as unavoidable adverse effects and irreplaceable or irretrievable commitments of resources, to be addressed. In the case history under discussion, these aspects were addressed as required.

Unavoidable Adverse Effects on the Environment. The unavoidable adverse effects of this highway project as proposed fall into roughly three sections: the use of some pieces of public lands, most notably parks and historical sites; effects on the quality of the environment with regard to air and noise; and, finally, effect upon the surroundings with respect to conservation and preservation of vegetation. With respect to adverse effects in regard to land, there were four identified specific sites that would be impacted if the project were constructed. The first two involved the taking of small pieces of land from two small neighborhood parks. The loss of land in each case was less than one-quarter of an acre from each park and, in the case of one of the parks, four and one-quarter acres could be made available to the park as a result of the abandonment of a section of one of the present interchanges. Of a more serious nature was the impact on the remaining two sites, the first of which was the local zoological gardens. From this site, the proposed or preferred alternate would require the taking of some three-and-a-half acres, presently undeveloped, in an area that would be used by the zoological garden if funds were available for expanding the present facilities. The fourth site was a local historical home, well over a century and a half in age, designed by Thomas Jefferson. While the home would not have been disturbed, it would have been necessary to have the highway unduly close to several of the outbuildings on the historical site. In addition to these land impacts, because the right-of-way would be expanded through existing residential and commercial property areas, the taking of any excess land would be quite expensive. Thus, the agency proposed taking a minimum width as a right-of-way to avoid displacing more people than necessary and to avoid as much of the cost of buying land sites and structures as possible. This would leave a sizeable number of residential structures and businesses, as well as the previously mentioned historical home and churches, unduly close to the edge of the right-of-way. Because of this proximity to the highway, these homes would be exposed to levels of noise and levels of air pollution which, if provision were not made for their reduction, would be above the acceptable limits if these structures were to be occupied. In order to reduce noise levels, the agency proposed that an earth berm barrier with a sound reflecting wall on the top be constructed in all areas where residential property was relatively near the edge of the right-of-way. With respect to the high levels of air pollution, the only statement that could be made was to predict that, because of the new federal regulations for air pollution control devices on automobiles, future levels of air pollution would be reduced below those which were deemed to be unacceptable.

The only other unavoidable adverse effect addressed in the statement was that of the loss of a substantial number of trees and strips of woodlands, which were providing supporting habitat for some small field animals, near several undeveloped sections of the highway.

Alternatives to the Proposed Action. As previously mentioned, alternatives were considered which include the do-nothing or null alternative. This was rejected on the basis that the increase in traffic would lead to increasingly adverse environmental effects from air and noise pollution, as well as a potential flooding condition along several sections of the highway. The mass transit option was rejected as being of such a nature that it would not develop adequate patronage to divert a sufficient number of vehicles from the highway, allowing the existing highway to carry predicted traffic at an acceptable level. Other options considered, but rejected, were a double-decking of the present highway and a second belt-line, appreciably removed from the existing belt-line, in order to avoid excessively serious displacement of a large number of people in already developed residential areas. In short, the impact statement recommended the rejection of all alternatives except the so-called preferred widening plan.

Short-Term Uses and Long-Term Productivity. The proposed improvement had as a goal the restoration of an adequate level of service for the highway, both immediately and for the long-term transportation needs as projected for the area. It was stated that the additional resources required for this improvement would be modest when compared to the benefits to the extensive areas served by the highway. The preferred plan was to take from their owners slightly less than 1,000 properties requiring the relocation of 835 families, the removal of some 62 businesses, and the dislocation of almost 900 jobs. These were recognized as significant impacts, but it was stated that the consequences of not improving the highway would be far greater. Without the proposed improvement, it was predicted that traffic service would continue to deteriorate, and the corridor of the highway would begin to play an adverse role in restricting the expansion of both residential and commercial properties in the suburban areas served by the highway. It was further predicted that failure to implement the project would contribute to a premature decline of the already existing residential, commercial, and industrial properties built along the expressway in the last 20 years. Such a decline would represent a substantial socioeconomic displacement, with its resulting loss of economic and human resources. The do-nothing alternative was predicted to have a substantial effect on the city by draining tax values from properties and by inhibiting the continued economic growth of the areas the highway served.

Irreplaceable or Irretrievable Commitments of Resources. Once the expressway was completed, the resources irretrievably committed would include the land which was transferred to highway use and, as a consequence, lost to any other kind of use. The commitment of materials and monies used in constructing the highway, of course, would be irreversible. Most of the irretrievable resources that would be used in the construction would be the materials required for the concrete pavement, which would consist of several million cubic yards of crushed stone from local limestone quarries, the accompanying cement, and the steel for the reinforcement of slabs and bridges. None of these commitments were considered to be serious, in light of the extensive quantities of these resources in the local area (with the exception of the steel required, which was reported to be readily available from outside sources). The large quantities of these materials to be used were acknowledged to probably cause temporary local shortages, but otherwise, no special deficiencies would be associated with their use for the construction of the highway. The most serious irretrievable commitment acknowledged was the additional loss of land that would be needed for the expanded right-of-way, which would, for all intents and purposes, commit the land to this use on a permanent basis and remove it from any other kind of productive or recreational use.

6-2.4 Summary

In this case history, it is important to point out several items for consideration. The first of these is the fulfillment of the requirements of NEPA by considering the various alternatives to the proposed action. All of the other items required by NEPA were covered, and the draft statement was circulated as required, with those comments received answered and included in the final impact statement. From this point of view, very little fault could be found with the statement prepared for this particular case. However, the primary fault with the project was one of inadequate public relations.

One of the first projects undertaken by the consultants engaged to prepare the impact statement for the highway expansion was a mail questionnaire and many door-to-door interviews with the people living in the residential neighborhoods immediately adjacent to the existing highway. The questionnaire was worded to elicit from the residents those effects from the existing highway which seemed to most disturb them. At the same time, each person was asked whether or not he favored improving the highway to accelerate the flow of traffic during the heavy rush-hour periods. As can readily be seen, these questions almost inevitably led the local residents to answer affirmatively. On the other hand, as has happened in so many other projects undertaken by a public agency, each resident assumed, in answering the interviewer or the questionnaire, that only beneficial impacts would be received and that there would be no adverse effects from the widening of the highway. The general public was not involved on a continuing basis in the planning, and was not contacted in this regard again until a series of public meetings were held. At these meetings, the preferred plan and the rejected alternatives were presented. In general, owners of parcels of property from which a segment was to be taken objected to the plans as proposed. This included not only the residential property owners, but also the owners of the commercial and church properties along the right-of-way, and the trustees of the historic home and the zoological garden.

This case definitely illustrates that the inclusion of the general public in assessments at the beginning and the end of a project is not sufficient. Had the general public been included on a continuing basis, many of the final objections could have been alleviated, either by changes incorporated in the plans or simply by making it clear to the property owners that this was the best plan available for the community as a whole. In this regard, the consultants and the agency planning this project were quite careless in their approach to public participation. As a result, the public clamor was of such magnitude that the project as proposed was completely rejected and, at present, is being reconsidered on a much smaller scale.

In stark contrast to the handling of the public participation in the highway planning in the case just outlined, was the method used by the Virginia Highway Department in its planning for the section of Interstate 66 in the Fairfax-Arlington community. In that case, the public was involved at a very early date and was kept continuously involved, not just informed, and actually was requested to participate in the planning at almost every stage of the project. Copies of the plans, feasibility studies, and other design information were made available in local libraries to citizen groups and to individuals for comment and participation at all stages of the planning.

Early in the planning, the department outlined a 7-phase study plan which included the following phases: Phase One—inventory;

Phase Two—evaluation criteria; Phase Three—identification of alternatives; Phase Four—evaluation and selection of feasible alternatives; Phase Five—detailed evaluation of the feasible alternatives; Phase Six—refinement of selected alternatives; and Phase Seven—conclusions. Phase Seven culminated in the preparation of an engineering report and a draft environmental impact statement. A community attitude survey was included as part of Phase One, the initial inventory, and from that point until the conclusion of the study, the public was continuously involved. In order that all current study facts were made available to all interested parties, a small eight page newspaper-like flyer was published by the Department, entitled "The I-66 Question: Fairfax-Arlington Transportation Alternates Study." This small circular-type paper was distributed to all who were interested in obtaining copies; actually, it was edited by the consulting engineer retained by the Virginia Highway Department to assess the impacts of the various proposed alternate highways. Not only was the public invited to comment, their participation was actively solicited. Study groups were formed in the various communities, and meetings were held at frequent intervals in order that any objections would be forthcoming at an early date. This "early warning" of objections gave the planners time to meet the objections or satisfactorily answer the questions while changes could still be incorporated in the plan, rather than after all planning and assessment was done (as was the situation in the case cited in this history). It cannot be emphasized too strongly that early and continuous public participation, particularly in controversial projects, will ultimately be beneficial to all concerned: the agency, the assessor, and the general public.

6-3. CASE TWO—INTERSTATE HIGHWAY

Another illustration can be given, of a case history for a highway project, which is similar in some respects to the previous case history, but differs in the assessment approach and in the impacts encountered because of the different type of topography and socioeconomic conditions. This case history is one of a project involving the upgrading to federal standards of a section of a midwestern interstate highway. This particular highway is a major link in the interstate highway system and connects the north central Midwest with the major routes to the southeast Gulf Coast and Florida. The project, in this case, involves a 45-mile long section of the highway, which was originally conceived and built as a toll road just prior to the inception of the Interstate and Defense Highway System in the United States. The highway was designed in the early nineteen fifties and was first opened to traffic about the middle of the nine-teen fifties. It was one of the first modern, four-lane, divided highways built in the Midwest and, as such, was designed before the federal interstate program was initiated. Because it preceded the interstate highway program, it obviously was not built to the same standards which were subsequently required with the passage of the Interstate Highway Act.

Almost from its inception, because of its geographic location as part of the major highway system to the Gulf Coast and Florida, it has carried a high density of traffic. Because of this heavy traffic projected for the highway by the consultants, the tolls were set relatively low for this type of highway, with the provision that when the bonds which were sold to build the highway were retired, the tolls would be removed and the highway would simply become a toll-free expressway.

When the Interstate System was established by Congress, this toll road's strategic location and capacity caused it to be designated a part of one of the major interstate routes, in spite of the fact that it was a toll road. According to the federal regulations of the Act, all interstate routes are required to be toll-free. This requirement, as mentioned, was mandatory when the outstanding bonds were re-tired. Over the years since its construction, and after the incorpo-ration of this toll section into the interstate system, a portion of the toll income has been set aside in a reserve fund to bring the highway to a high state of repair at the time the bonds were retired and the tolls removed. The funds accumulated, however, were not sufficient to bring the road to present day interstate standards, and thus, it will be required that these funds be supplemented by addi-tional state and federal funds for the upgrading of this section of the interstate. Because of the requirement for the federal funds to bring the road to acceptable interstate standards, the project falls under the aegis of NEPA, and an Environmental Impact Statement is required.

The critical problems which must be solved by this project include inadequate lanes for future projected traffic, correction of the narrow 20-foot median, the lack of roadside recovery areas for out of control vehicles, fixed obstructions too near the travel way, substandard and inadequate interchanges, and several long uphill grades, causing trucks to move slowly and thus impede traffic flow.

The project, as envisioned in the impact statement, would entail the addition of one lane in each direction for a distance of some 30 miles. This part is the portion of the highway which passes through the predominantly rural and small community portion of the project. In the remaining 15 miles of the project, the surroundings are more heavily urban, with the highway section actually terminating in a heavily residential portion of a major midwest metropolitan community. In this section it is proposed to add two lanes in either direction, in addition to the inclusion of new frontage roads.

6-3.1 Environmental Setting Without the Project

This case is somewhat unusual in that the null alternative is, for all practical purposes, not available to the agency. The highway as it exists is designated as a part of the interstate system, but is not up to interstate standards as required by the federal act. Thus, the null alternative, which, in this case, would simply be the removing of the tolls and opening the highway to toll-free traffic, would not be acceptable to the Federal Highway Administration, and would, in fact, be a violation of existing federal safety standards for a segment of the interstate system. The highway, by requirement of the law, must be improved, with the addition of the required safety features, whether or not provisions are made for relieving the traffic congestion and adapting the highway for the increased traffic capacity which will be necessary when the tolls are removed.

Essentially, the highway travels through four extremely different conditions of topography and geography. The beginning and end of the segment of the highway considered in this project are both in urban, well-developed, residential and commercial areas. The second type of topography is that of those segments in which the highway passes through residential areas that are very small and, in many cases, unincorporated communities. The remainder of the highway passes through either farm lands or undeveloped public lands. Of the 45-mile improvement proposed in this project, approximately 15 miles will pass through heavily developed urban areas, two to three miles will be through sparsely developed urban areas, five miles through public lands, primarily forest areas, with the remainder passing through privately-owned farmland.

A somewhat unique topographical feature of the existing highway is that, for a major portion of its length, it is bounded closely on one side by a major mainline railroad. In other stretches of the highway, it is bounded by a major stream, with the railroad bordering the stream on the side away from the highway. These two features limit, in many respects, the alternatives available for widening or moving the highway right-of-way in the direction of the stream or railroad.

The major environmental problems with the highway, as it currently exists, are those which exist along the portions in the highly developed urban areas. The pollutants of excessive noxious gases from the vehicles, and excessive noise from the trucks on the heavily travelled highway are the major concerns.

From the point of view of safety, and also from the concern in regard to the lack of traffic carrying capacity, the highway, without the project, is grossly inadequate. At many times during the year, particularly during holiday weekends, traffic is almost at a standstill; bumper to bumper for miles in several stretches of the project. Since the construction of the highway, drivers have quite frequently lost control of their vehicles for one reason or another, crossed the narrow unguarded median, and crashed head on into the oncoming lane of traffic. As a portion of the interstate system, the segment to be improved by this project has one of the higher fatality rates per mile of the entire 42,500-mile Federal Interstate Highway System.

From an archeological viewpoint, the highway is built in an area which generally has been occupied by prehistoric groups, but specific information on any particular sites of historical or archeological significance immediately adjacent to the project is not precisely known. No prehistoric man-made structures, such as walls, earth embankments, or mounds were found in the project area. However, other evidence of prehistoric activity was found on the fringes of the project area and others just outside the area. Earlier investigations had shown finds of artifacts from several localities along and nearby the highway. Such artifacts as projectile points, hammer stones, pecking pebbles, pottery, and tools made from bone, antlers, and shells were found at various times in the past.

Sociologically, the residential areas vary from a small number of farm houses, which in many cases have been occupied by three or four generations of the same family, to the relatively recent, heavily developed, residential subdivisions in the urban areas. Most of the heavy residential areas at the two terminals of the project are less than ten years old as subdivision developments, with most of the homes falling in the $15-25,000 class. As a result of the subdivision developments, many homes were built immediately adjacent to the existing right-of-way and, in many cases, are less than 100 feet from the existing edge of the heavily travelled highway. Thus, many of the residents receive excessive amounts of noise from the highway as it presently exists.

Fig. 6-1   Deep rock cuts limit the width of right-of-way and leave steep grades, resulting in traffic slowdown.

The topography in the project area varies widely. Very little of the terrain within the project area rises above 1,000 feet above sea level in altitude, with the major part lying at altitudes of 500–600 feet. The northern portions are on the alluvial plains and are relatively flat. South of these low land areas, the highway passes through sections of erodible limestones and shales, becoming gently rolling. In the center portion, the highway passes through several sections of steeply sloping topography, which has required deep cuts into the surrounding terrain, at the same time leaving steep grades on the highway. These steep grades have been responsible for much of the slowdown of traffic as the heavy trucks have attempted to negotiate these grades. On the southern terminal, the topography again returns to more gently sloping terrain, with the highway being restricted to smaller, more easily negotiable slopes.

6-3.2 Impacts of the Proposed Action

As previously mentioned, the null alternative was not a viable option to the agency preparing the action for this particular project. Other alternatives, however, were considered, the first of which was a complete relocation of the highway. However, because of an extensive military reservation on one side of the area, and a rather sizable forest preserve and high elevation topography on the other side, the present corridor seemed to the agency to be the only feasible route. There were four alternates considered on the particular route. Basically, these consisted of widening to one side or the other, splitting the widening with a portion on one side and the remainder on the other side, or using the median as a paved additional lane with a concrete guardrail down the center. The preferred alternative consisted of a combination of widening on one side or the other, with the choice of which side to be widened depending primarily upon the topography and the type of adjacent land ownership in the particular case. The exception to this pattern was the very heavily developed residential area in the northernmost six to eight miles, in which it was proposed that a concrete median barrier be constructed, with the present median being used for one additional lane in each direction and a second additional lane to be added on either side. To develop this last option, it would also be necessary to construct a collector-distributor system parallel to the highway to facilitate the entrance and exit of the local traffic in these heavily developed areas.

The environmental impacts of the project as detailed by the consultant consist of both positive and negative effects. The positive impacts would be the replacement of a crooked, hazardous, and ineffecient forty-five mile highway section with a modern six or eight-lane facility. The project would have vastly improved highway safety potential, as well as a reduction in travel time, because of the improved flow of traffic. Improvement of the interchanges would expedite the flow of traffic entering and exiting from the highway and, as such, would relieve some traffic congestion in the adjacent developed areas. The air quality in the traffic corridor was predicted to be improved as a result of the increased average traffic speed, as well as the projected improvement in air pollution control devices on the vehicles. The proposed improvements would entail the lowering of the ruling grades in several areas of the project, which would alleviate much of the truck traffic slowdowns. This would also enhance the air quality in the corridor. The proposed widening was also found to be the preferred action from the standpoint of impacting on historical sites.

Many negative impacts were predicted and admitted in the EIS. Among the negative impacts would be the dislocation of 35 households and 21 non-residential properties. These dislocations would, of course, involve the trauma of moving which is always present when the move is involuntary and caused by conditions over which the individual has no control. Adjustment to new social relationships is forced in many cases, causing additional sociological impact. Some land would be removed from the existing tax roles, with a loss of revenue as a result of the taking of the various residential and commercial properties in the areas to be widened. A small amount of land would be lost for agricultural purposes. Some aesthetic loss and interference with neighborhood tranquility would be suffered, as well as the loss of some tree screens, which would be removed as the right-of-way was widened. Some soil erosion and sedimentation in nearby streams was predicted as the result of excavation and embankment construction.

Perhaps one of the most serious impacts would be the increase in existing noise levels, as a result of both the heavier traffic flow and the increased average speed. Noise levels were predicted to increase from two to four decibels as far away as 500 feet from the highway. In the heavily developed residential areas, the L_{10} noise level (that level existing an average of 10% of the time) would be raised above the acceptable 70 decibels for 30 to 35 homes in the rural and sparsely developed areas. In the heavily developed residential portions, the noise levels would be raised above the acceptable 70 decibels for a much larger number of homes; thus, extensive sound reduction barriers will be required as a part of the project.

There will be some terrestrial habitat reduction, due to the removal of tree stands and marsh areas, as well as impacts on the aquatic environment, due to the increased siltation and sedimentation in adjacent streams. An increased amount of chemical pollutants can be expected from the various chemicals which would be spread on the highway in times of freezing rains and snow.

Fig. 6-2 Residential developments close to the highway will result in severe noise impacts unless sound barriers are used.

The expected air quality levels would exceed federal standards at least once a year, and if any delays are forthcoming in the implementation of federal automotive emission standards, greater air pollution levels can be expected.

Even though no historically significant archeological sites were found in the right-of-way during the sampling program, archeological sites are a non-renewal resource. Therefore, any positive archeological gain is an impossibility, and the only possible impact, if any, from the construction and widening of the 45-mile highway project would, of necessity, be a negative one. In summary, the most serious impacts would seem to be the dislocation and forced relocation of the residential and commercial properties, and the probable increased levels of the noise for the residents adjacent to the highway who would not be forced to relocate. In this latter case, the seriousness of the impact, to a large extent, will depend upon the successful design and construction of noise attenuating barriers along the highway right-of-way.

Other minor short-term impacts would occur during the construction of the project itself. These short-term impacts would include increased noise levels, additional dust and air pollution from the construction vehicles, and the necessary detours which would be involved in order to maintain the flow of traffic during the construction period.

6-3.3 Other Aspects of NEPA Relative to the Project

As in the previous case, with the requirement of an EIS, it was necessary to address the other subjects which are specifically required by NEPA.

Unavoidable Adverse Effects on the Environment. Among the unavoidable adverse effects would be the increased traffic flow through the existing developed areas, with the resulting increased levels of traffic noise. The previously mentioned sociological dislocations would also be unavoidable since, in some areas, residential development is immediately adjacent to the highway right-of-way, so that any widening, however slight, would require the demolition of at least some residential property.

Since much of the length of the project is through undeveloped rural areas and farmlands, an abundant amount of wildlife is found in these areas adjacent to the highway. The widening required through these areas would entail the loss of some tree screens and marsh lands which will cause disruption in the habitat of some of these wildlife species.

In several cases, rural land owners would find themselves cut off from the use of some portion of their lands. This would entail the development of new access routes to some sections of isolated farmland, which could be a severe inconvenience from the standpoint of machinery movement and harvesting, for some farmers in the area.

The public acquisition of land for the project would of course involve the removal of such lands from the tax base of the local taxing jurisdiction. Based upon current tax rates in the area, it was estimated that the annual tax loss would exceed $10,000 per year, with more than 85% of this loss associated with improvements of the interchanges in the urban areas.

Relationship Between Local Short-Term Uses of Man's Environment and the Maintenance and Enhancement of Long-Term Productivity. From the standpoint of the short-term use of the environment, the impacts from this project would be relatively small. The major impacts would be the short-term additional noise and air pollution levels, some inconvenience from traffic re-routing, and an increase in the level of noise and dust from construction machinery. Some soil erosion with resulting muddying conditions, and sedimentation in adjacent streams, would be present should heavy rains be encountered during the construction period. However, the long-term improvement of the highway, with its resulting more convenient and safer flow of traffic, should more than offset the short-term impacts. When the project is complete, the increased levels of noise, and soil erosion and the resulting sedimentation, should be reduced or eliminated. While some increased noise levels will remain, the long-term levels will be less than those found during the construction period, with the presence of heavy construction machinery. Upon completion of the project, landscaping and plantings will be undertaken, and while some period will elapse before full growth is obtained, tree screens and other shrubbery for the smaller wildlife will eventually be restored. It is, however, improbable that those marshlands which will be required to be drained to facilitate the construction of the project will be restored after the construction is complete.

Irreplaceable and Irretrievable Commitments of Resources. The major commitment of resources which will be irreplaceable or irretrievable would be the land which will be committed to the highway right-of-way. This land will of course be lost for other purposes. There will be an irretrievable commitment of mineral resources, the sand and aggregate and cements used in the construction of the highway pavement, but since these are in abundant supply in the project area, this is not a serious commitment. As previously mentioned, archeological sites are an irreplaceable resource; should any sites be destroyed during the construction of the project, this would be an irreplaceable commitment. No irretrievable or irreplaceable commitments of fauna or flora would be created during the construction of this facility. Most of the flora and fauna that would be destroyed could be replaced after the project was completed. The preferred alternate plan specifically avoided impacting virgin forested areas, the loss of which would be irreplaceable. No serious dislocations of streams or lake areas were anticipated, and no resources would be irretrievably lost in this respect.

6-3.4 Summary

In this project, the negative impacts consist primarily of the dislocation of people in the urban areas and the increased air and noise pollution levels in these same areas. The project itself is almost compulsory, in that it is a part of an existing interstate highway which does not meet acceptable standards which have been imposed by law. It thus becomes imperative that the local agency proceed with one of the alternatives which will bring the highway to acceptable federal standards in regard to traffic safety and traffic capacity.

The most serious deficiency in this case is the complete absence of any public involvement in either the preparation of the environmental assessment or in the project planning. The actual assessment of the environmental impact was accomplished by a well structured interdisciplinary team of scientists and engineers. However, it was obvious from the draft impact statement that much of the effort was segmented and fragmented, with no public involvement in any aspect. It is, of course, required by NEPA that the EIS be circulated to interested citizen groups, and that public hearings be held at which the preferred plan must be presented, but at which the alternate plans must also be presented. Since the public and interested citizen environmental groups were not involved in any of the decision making prior to the release of the draft environmental statement, it is highly likely that much public disagreement will be found to the proposed plans of the agency; many parts of the project may very well be wholly unacceptable to some or all citizen groups. This is yet another case where adequate public relations and public involvement in the early stages of the planning would probably have prevented a great deal of public hue and outcry which could be of such a magnitude that it would lead to some type of legal action against the project. In view of the many court decisions, referred to in detail in the earlier chapters, it seems only likely that some legal action will be considered for almost every project of this nature in which complete and frank public involvement is not undertaken from the project beginning and continued throughout the entire project. Dislocation of families from residential areas, cutting of farmlands into segments, and increasing the levels of noise and air pollution on the residential and commercial property not taken during the construction of the project, are all areas which, in the past, have been found to lead to great public upset and the threat of legal action against projects.

In cases such as the one just discussed, the citizens to be impacted are rather well delineated. It thus seems only logical that these intimately involved groups, as well as concerned officials, should be included in the project planning at an early date. A well developed public relations program, continuous throughout the project, which relates to the public the data and projected impacts as they are developed will, in most cases, allay fears and eliminate many legal actions of the type which have caused the abandonment of many well-engineered projects that unfortunately have had poor public relations. In this respect, Congress and the courts have been quite clear; the planning of any project must not proceed without public participation and involvement.

The other unique feature of this case is the necessity, by law, to do something. In this situation, there would seem to be no "null" alternative that the agency can consider.

6-4. CASE THREE—WATER RESOURCES PLANNING

The third case to be presented in this chapter describes an application of environmental assessment methodology to predict impacts associated with the development of a water resources management plan for a major metropolitan area. Generally, assessment activities associated with water resources projects fall into one of several categories: monitoring of water quality to characterize the existing environment without the proposed project or activity; prediction of impacts from a proposed water resources management activity on the basis of a predictive model of the environment, the proposed action and its associated generation of pollutants; and the evaluation of the impacts identified on the basis of model studies, with regard to direct and indirect effects, short-term and long-term effects, secondary effects, and socio-political effects, as well as effects on the physical and biological environment.

6-4.1 Background Information

The background information in this particular case is based upon the preparation of a water resources management plan for a metropolitan area located along a major river in the Midwest. A consulting engineering firm had been selected by local planning and development agency personnel to develop a 201 Facilities Plan, according to the guidelines of the Environmental Protection Agency, for each of the three counties which compose the major metropolitan area. The requirements of the EPA for completion of the planned development included the preparation of an environmental assessment of the effects of implementation of the proposed water resources management plan. Because of the unique sequence of events in this particular case, the preparation of the impact assessment served to summarize much of the work which had been done in developing the facilities plan.

During the early stages of the development of the 201 Facilities Plan, the physical system in the three-county area had been characterized through the preparation of a comprehensive inventory and basic description of the physical characteristics of the area. The geology, soils, topography, geography, climate, and hydrology of the area were analyzed as the initial step in the development of the water resources management plan. Since this information was to be used in developing a water resources facilities plan, particular attention was devoted to identifying characteristics of the region which would have particular bearing on the quantity and quality of water in the regional water system of which the study area subsystem was a part. For example, the low permeability of surface layers of soil and rock in some portions of the study area precluded the use of septic tanks for wastewater management. In contrast, in other areas of the metropolitan region, the presence of cavernous limestone beneath the surficial soil layers presented particular problems of groundwater, surface water, and wastewater management.

The principal feature of the water flow system in the study region was the major river mentioned previously. This river was flowing in a valley filled with granular materials. The bedrock surface had been eroded to a much greater depth than the present land surface, during previous geological periods. During subsequent times, when the river level rose, deep layers of clean sand and gravel were deposited over the bedrock surface. In recent times, layers of silt and clay were deposited by the modern river during periods of flooding. Characteristically, the river is bordered by natural levees of rather silty material. Adjacent to the levees, on the side away from the river, are found floodplain areas containing fine-grained soils such as silts and clays, which were deposited by flood waters trapped in low-lying areas. Since the river flows in a bedrock valley which is filled with pervious sand and gravel, the groundwater flow system in the sand and gravel reflects changes in the river flow regime very closely. In many locations, an excellent hydraulic connection exists between the river and the sand and gravel subsoil layers. The groundwater in the granular soils adjacent to the river constitutes a major water resource in the study area. Significant amounts of groundwater are not found in other localities and the granular valley filling represents the most important groundwater reservoir in the entire area.

Most of the tributary streams which flow into the river in the study area are very much smaller than the main stream. These streams characteristically have seven-day, ten-year minimum flows of nearly zero; this extremely low dry weather flow greatly limits their capacity to assimilate wastewaters. This lack of assimilative capacity in many of the surface streams throughout the study area presents one of the most significant water resources management problems for planners in this metropolitan region. Much of the effort put into the development of the 201 Facilities Plan for this region was directed toward developing a system for treatment of wastewaters with subsequent discharge of the treated waters into the principal stream running through the area.

As a part of the development of the 201 Facilities Plan, the consulting engineers employed by the regional planning agency studied the social, economic, and land-use characteristics of the metropolitan region. Population estimates were made for the current year and these estimates were projected 20 years into the future. The data for these projections were furnished partly by the EPA and partly by the planning agency for the metropolitan area under investigation. In order to prepare estimates of current and projected quantities and types of industrial wastewaters in the study region, an industrial inventory for the entire region was prepared. Finally since the guidelines of the EPA, for 201 Facilities Planning, stipulated that these plans should be based upon acceptable or approved land use plans, all local planning agencies in the entire metropolitan region were contacted by the consultants developing the 201 Facilities Plan. These agencies were asked to develop a comprehensive land use plan, in conjunction with the consulting engineers, for the three counties involved in the study effort. Land use in the metropolitan region was characterized through collecting comprehensive data, including the future distribution of residential population in the study area and the current and future location and size of industrial facilities. Projections for a 20-year period were based upon the capability of the land units throughout the study area to support certain uses, and a rational projection of development and growth of existing communities within the three-county region.

6-4.2   The Environment of the Project Site

In this particular case, the planning involved in developing the 201 Facilities Plan included a characterization of the environment in the study area, as outlined in previous paragraphs. Additionally, a special effort was made to identify and characterize the aquatic environment, since the purpose of the proposed study was the management of water resources in this area. An effort was made to identify and characterize all existing or permitted sources of wastewater in the three counties under-study. These wastewater sources were characterized in three major divisions: sources producing more than 100,000 gallons per day; sources producing from 10,000 to 100,000 gallons per day; and sources producing less than 10,000 gallons per day. Approximately two-thirds of the identified wastewater sources produced less than 10,000 gallons per day, as might be expected for a metropolitan area in the United States at the present time. More than one-half of the total wastewater generated in the study area was generated in dwelling units (residences, apartments, and mobile home facilities); these residential wastewaters were processed in municipal wastewater treatment facilities, and in small, package treatment plants. About 40% of the total flow of wastewater was industrial wastewater. This amount of wastewater was processed in part by the industries themselves and in part in municipal wastewater treatment facilities. Commercial, recreation, institutional, and miscellaneous sources of wastewater accounted for only about 2% of the total flow of wastewater in the study region.

The overall quality of the aquatic environment in the study region was evaluated by monitoring the water quality in surface streams and in groundwater reserves within the three counties in the study region. Numerous samples were obtained in surface streams and in wells to characterize the water resources of the region. More than 100 sampling stations were established in the three counties under study in order to determine the existing water quality in the region, and to identify sampling locations which would furnish the most representative samples of water for analysis and evaluation in predicting water quality in the future. The results of the sampling and testing activity indicated that many of the streams in the three-county area exhibited high bacterial levels showing probable fecal contamination. Additional parameters were monitored during the preparation of the facilities plan. Evaluation of the testing results with respect to these parameters showed that the levels of dissolved oxygen, pH, and total dissolved solids in the obtained samples did not conform to the criteria for public drinking water supplies as promulgated by the United States Public Health Service. Sampling also was carried out at various locations on the major river flowing through the study area. Analysis and testing of these samples indicated that the quality of the water in the major river was relatively poor. The poor quality of the river water was felt to be a reflection of the poor state of wastewater treatment throughout the study region. Analysis and testing of samples taken in two of the largest tributary streams in the study region showed consistant results and indicated that the quality of the water in these streams was extremely poor. These streams were characterized by insufficient dissolved oxygen content, the presence of fecal and total coliform bacteria in excess of federal standards, and dissolved solids in concentrations far in excess of applicable standards.

During the course of the development of the facilities plan for the metropolitan region, a model of the surface streams in the study area was prepared. This model was intended to allow a determination of the assimilative capacity of the major receiving streams in the area, on the basis of the water quality sampling and testing which had been completed as a part of the plan development. In a preliminary investigation, the model was used to evaluate water quality in the streams of the region. This preliminary investigation indicated that the major river in the study area was the only stream that could accept significant amounts of wastewater, if the wastewaters had been subjected to only secondary wastewater treatment. Many of the small tributary streams in the study region were found to have very low assimilative capacities because of their insignificant dry weather flows. These streams could accept only wastewaters which had been subjected to a high level of treatment.

In order to characterize the total water environment in the study region, the developers of the 201 Facilities Plan analyzed the existing capability for wastewater treatment in the region. Three of the largest wastewater treatment facilities in the metropolitan area included only primary treatment of wastewaters. One of these facilities, the largest plant in the metropolitan region, was to be expanded in the near future to include secondary treatment facilities. Some of the smaller treatment plants in the study area were equipped to provide secondary treatment. However, these plants were found to be discharging into streams with very low assimilative capacity; capacity so low that only relatively pure effluent could be safely received. In other words, a very high degree of treatment would be necessary in order to introduce wastewaters into these streams without posing an environmental hazard to persons in the study area. A large number of the package treatment plants in the study region had been constructed to serve small residential subdivisions. Less than 10% of the package treatment plants in the study region were found to be equipped to provide tertiary or advanced treatment, although most of these plants were found to be discharging effluent into streams with very low assimilative capacities. In essence, most of the streams into which package treatment plant effluents were being discharged were very sensitive to the introduction of such effluents.

6-4.3   Other Aspects of the Environment

In order to determine the potential for development of a systematic wastewater management plan in the study region, the consulting engineers on this project analyzed the political, jurisdictional, and administrative conditions of the governmental bodies of the three counties under study and of the cities located within those counties. Since the Water Pollution Control Act of 1972 (as amended) required consolidation of existing treatment facilities wherever possible, with subsequent development of regional wastewater management agencies, the developers of the 201 Facilities Plan recommended the establishment of county-wide sewer districts for each of the three counties under investigation. The planners felt that development of such county-wide sewer districts would help to eliminate point sources of water pollution, would help to improve overall water quality, and would assist in obtaining economy of scale in the design and construction of treatment plants and associated facilities. It was necessary to determine the fiscal capabilities of the governmental units in the three counties of the study area in order to undertake design and construction of wastewater treatment facilities. This study of available financial resources included a determination of the availability of federal funds for water quality management activities. Also, revenue-sharing funds and community development funds available from the federal government were investigated. State programs which were available to provide matching funds to assist small communities in water quality management planning were studied, too. All these potential sources of financial backing were analyzed and evaluated in order to determine the total financial capabilities for water resources activity in the three counties of the study region.

6-4.4 Alternatives to the Proposed Action

As part of the development of the 201 Facilities Plan, the consulting engineers for the regional planning agency in this area gathered all of the information described in the preceding paragraphs and employed systematic optimization techniques to determine all viable alternatives for wastewater management in the study region and to identify the wastewater management alternatives which would produce the lowest capital and operating expenses. Alternative sewer system configurations, alternative treatment facility locations, alternative degrees of treatment, alternative degrees of correction or treatment of infiltration and inflow, and other alternative factors were included in this optimization effort. As a result of this phase of the study, more than 20 alternative schemes for wastewater management were developed for the three counties in the metropolitan region. On the basis of an evaluation of these alternatives, it was determined that the total flow of water through the study area and, in particular, the flow of wastewater in the region could be reduced by reducing the infiltration and inflow into existing and planned sewer systems in the region. The possible sewer system configurations were evaluated on the basis of the identified trends in land use and on the basis of the locations of sewer lines required for the consolidation of treatment facilities in the region. The various viable treatment alternatives were evaluated; this study indicated that all new plants located to discharge into the major river in the study area would require secondary treatment capabilities. In contrast, any small plants built on tributary streams in the study region would require advanced wastewater treatment capabilities, or would require application of generated effluents to the land surface, instead of discharge of effluents into receiving waters. In this evaluation phase, it was determined that reuse of wastewater and land application of effluents were not feasible for technical, economic, and environmental reasons. Because of the existing poor water quality of many of the small streams in the area, and because of the very limited assimilative capabilities of these streams, it was decided that a regional wastewater treatment system would be very advantageous. Such a regional system also would be advantageous with respect to the economies of scale associated with centralizing of treatment facilities. Also, development of a large treatment system would furnish possible capabilities to extend wastewater services into remote areas of the three counties in the metropolitan region which were not served by sewers or package treatment plants.

6-4.5 Public Participation

An important part of any overall environmental assessment is participation by the public in the evaluation and planning process. In this particular instance, public participation was achieved in the planning process, and much of the difficulty in preparing a comprehensive assessment was alleviated through this public input into the development of the basic facilities plan. A citizens' evaluation committee was created to furnish input into the development of the management plan. This committee included representatives of various public and private organizations. One of the principal tasks of this committee was evaluation of the alternative wastewater management plans which had been identified through the systematic optimization study conducted in an earlier phase of the plan development. The advisory committee spent more than 50 hours in conference in identifying potential problem areas and in evaluating the proposed alternatives. Additionally, members of the advisory committee also proposed modifications to several of the existing alternative plans. More than a dozen separate plans had been identified as the most cost effective alternatives for the two counties located to the north of the major river in the metropolitan region. Eight distinct plans had been identified as viable alternatives for the area to the south of the river. Included in these plans were the possible locations of plants at eight different sites north of the river and at six sites south of the river. A variety of different degrees of wastewater treatment at each of the plants were included in each of the proposed alternative plans. A further modification of the developed plans, which constituted additional alternatives, was the correction of inflow and infiltration outside combined sewers, the treatment of infiltration and inflow outside combined sewers, and the correction of infiltration and inflow for combined sewers with the addition of detention basins for the combined sewers.

As a result of considering all of these possible modifications in the basic alternative plans, the citizens' committee considered a very large number of possible alternatives. In evaluating these alternatives, the advisory committee developed a matrix of parameters for use in judging each particular wastewater management scheme. The parameters which were used in this evaluation process included: overall effect on water quality; impact on land use; environmental effects; financial capabilities to complete each of the plans; the reliability/flexibility/operability of each of the plans; capabilities of governmental units to manage the facilities and to enforce regulations; project timing for each of the alternatives; success in eliminating existing wastewater sources; coverage of the study area with services; capabilities to upgrade existing facilities; total project costs; energy availability and conservation; availability of total system resources; capabilities to recover and reuse natural resources; and compatibility with water supply requirements for the metropolitan area. All of these parameters were given quantitative values by the members of the study committee. The committee members used a scheme similar to the Battelle system mentioned in Chapter 5. A total effectiveness value (a number) was obtained for each alternative plan. The cost effectiveness of each of the plans was determined. On the basis of all these determinations, the advisory committee selected the best plan for each of the three counties in the study region. Subsequently, a public hearing was held to describe the selected plans and to obtain citizen comment. No objections to the plans were presented at the public meetings.

6-4.6 Environmental Assessment

At this stage in the development of the overall facilities plan, preparation of an environmental assessment of the proposed plan was undertaken. The alternatives which had been identified as feasible by the project initiators and by the citizens evaluation committee also were to be evaluated with respect to overall environmental impact. The task of preparing the environmental assessment was greatly simplified by the compilation of environmental data which had been made in the course of developing a characterization of the region as an initial phase in the development of the 201 Facilities Plan. This description of the physical system and the man-made environment in the study area was sufficient for preparation of an overall description of the environment without the proposed wastewater treatment facilities.

The proposed wastewater treatment plan was described in detail for the system which had been formulated on the basis of the initial technical, economic, and environmental analyses. The location of the wastewater treatment facilities and the configuration of the proposed sewer systems in the preferred alternative plan were described. It was possible, in the environmental assessment, to give a complete proposed schedule for construction and completion of the treatment facilities. Also, very importantly, the relationships among the proposed facilities and the other proposed water quality management efforts which were planned for the study region were described in the environmental assessment. This capability to describe the relationships between the regional plans was a very desirable consequence of the previous development of the 201 Facilities Plan. The characterization of the environment in the study area, and the projections for future development in the study area, without the proposed treatment facilities, as developed for the 201 Facilities Plan, permitted the consulting engineers to develop a very comprehensive assessment of the impact of the null alternative.

During the preparation of the facilities plan, and during the accumulation of much of the information describing the environment in the study region, a very large number of maps, photographs, and renderings had been prepared. Many of these graphical aids had been prepared for use in describing the physical system and possible alternative plans to the community advisory committee. These graphic aids were very useful in the description of the existing environment and the proposed projects which were necessary for the environmental assessment study. Each of the alternative water quality management plans was described in significant detail. This capability to describe each of the alternatives in detail was a result of the detailed evaluation of alternatives which had been completed by the community advisory committee.

This evaluation and description of the existing environment and the proposed plans was facilitated by dividing the study region into selected water quality management areas on the basis of drainage basin boundaries and the boundaries of groundwater recharge areas within the three counties of the metropolitan region. The guidelines of the Council on Environmental Quality required a description of the relationship of the proposed water quality management plan with current and proposed land use planning in the three county area. This relationship was analyzed and described. Expected future development of the area was described on the basis of the existing land use patterns in the region and on the basis of identified trends. Future changes in land use and growth in population were used to establish future probable water demands and resultant wastewater discharge volumes. The land use planning and population growth characteristics of the region were incorporated into the assessment study analysis in order to provide a clear understanding of the future demands on the water resources of the study region. In the evaluation of suggested alternatives by the community advisory committee, the committee considered the compatibility of the 201 Facilities Plan with existing and projected land uses to be a very significant relationship. The committee had used 15 criteria in evaluating the various alternative management plans. Compatibility of the proposed plan with existing and projected land uses was considered to be the second most important of the 15 evaluation criteria, exceeded only by overall water quality. This emphasis on compatibility of the facilities plan with existing and future land uses stressed the possible positive and negative impacts which could result from the design, construction, and operation of the various proposed sewer configurations and treatment facilities.

6-4.7 Assessment Methodology

In evaluating the environmental impact of the proposed water quality management plan and the alternative plans, the consulting engineers assigned to this task utilized a system which was very similar to the system proposed by personnel of the Battelle Memorial Laboratories, as described in Chapter 5. Four general categories of impact were identified: ecological impacts; impacts on physical environment; socio-economic impacts; and aesthetic impacts. With regard to the ecological environment, four environmental parameters were considered in judging impacts: effects on aquatic flora; effects on terrestrial flora; effects on aquatic fauna; and effects on terrestrial fauna. The impacts on the physical system were quantified with respect to impacts on water quality, land quality, air quality, hydrology, geology, and noise levels. The socio-economic parameters which were used in identifying and evaluating socio-economic impacts included existing and future land use, economic development in the study region, archeological resources, historical sites, recreation facilities, existing and planned transportation facilities, and cultural features in the study area. One of the most difficult categories of impact to assess in this investigation, as in all assessment studies, was the impact on the aesthetic quality of the study area. In this particular investigation, the consultants studied the physical factors in the region, the spatial composition of the various physical factors, the man-made objects in the three county region, and any unique aesthetic factors located in the area. Each of the parameters cited above was assigned a number of environmental quality points. A mechanism for obtaining these environmental quality points, similar to the method for obtaining parameter importance units in the Battelle system, was used to quantify the impacts of the various proposed plans and alternatives for each of the four categories of environmental quality. With respect to the ecological environment, a total of 250 environmental quality points were assigned to the four parameters describing its quality. Physical environmental quality, represented by the six parameters previously mentioned, was assigned 500 environmental quality points. The seven socio-economic factors mentioned above were assigned a total of 250 environmental quality points, and the aesthetic factors were assigned 100 environmental quality points. Thus a total of 1,000 environmental quality units were distributed among the 21 different parameters which had been used to describe the quality of the existing environment in the study region. In determining the degree of impact of the proposed plan, the consultants used a rating from −5 to +5 to evaluate the effect on any given parameter of the operations included in the management plan. The rating for each of the factors was multiplied by the number of environmental quality points assigned to that factor, in order to determine an overall rating for each alternative plan.

In accordance with the requirements of the National Environmental Policy Act, the consultants determined the environmental impact of the null alternative. The data which had been accumulated in the previous portions of the plan development were utilized to describe the existing treatment facilities, the existing water quality in the project region, the predicted trends in population and economic growth, and the characteristics of the physical environment in the study region. After ratings of impact for each of the environmental parameters mentioned above were obtained, it was found that the rating for the null alternative indicated that this alternative would have a very significant detrimental impact on water quality in the region. This impact was found to be most severe in a heavily industrialized portion of the area to the south of the river. In contrast, the proposed management plan was found to have a very significant predicted beneficial effect on the water quality in the area. In terms of the evaluation system which had been proposed by the consultants, the proposed management plan had a total environmental quality score in excess of 1,200, while the null alternative had an overall score of slightly less than −100.

The potential impacts of the proposed plan included improvement in the overall water quality of the study region, as well as decreases in the possible levels of air pollution in the region, through elimination of sludge incineration at several small water treatment plants located throughout the area. Additionally, construction of the centralized treatment facilities was predicted to have a beneficial effect on total water quality, through elimination of many of the small treatment facilities which were discharging effluents into intermittent and ephemeral streams incapable of assimilating those discharges.

The detrimental impacts associated with the proposed plan included disruption of the physical system in the study area during construction, and operation of the treatment facilities and the sewer systems included in the proposed plan. The consulting engineers engaged in this assessment enumerated these impacts in a number of different categories. For example, the known archaeological sites in the study area were investigated, and it was discovered that several sites would be disturbed to a certain extent by the proposed construction of treatment facilities and sewer lines. Recreational areas, including parks, golf courses, and other facilities, also were found to be affected to some degree by the proposed construction activities. The proposed facilities and sewer lines were depicted on maps in order to establish their relative location with respect to archaeological and historical features. On the basis of a comprehensive evaluation of possible effects on these resources, it was shown that the overall impact of the proposed construction on cultural features was minimal.

Another example of the detailed studies in the assessment was the review and evaluation of impacts for the wooded areas in the study region. The principal impact in these areas resulted from disruption of the physical environment through construction of sewer lines. However, the construction activities associated with the proposed plan were predicted to have only slight short-term impacts on the terrestrial flora and fauna in the area, although all terrestrial flora in the path of the construction activities would be destroyed. It was decided to mitigate this impact by replacing, wherever possible, the destroyed terrestrial flora. No significant total habitats were predicted to be significantly altered by the proposed construction facilities. In general, the effects on the ecological environment were predicted to be rather temporary.

An interesting development in this environmental assessment occurred with regard to historical sites. As mentioned previously, a number of archaeological sites had been found in the study area. Likewise, a number of historical sites were found in the region, and the possibility of disruption of these sites by construction and operation activities was evaluated by the consulting engineers. Several of these sites were found to be listed on the National Register of Historical Sites. Other sites were evaluated and found to be eligible for registration, although they were not to be found on the Register. In evaluating total impact, equal importance was attached to eligible and registered sites. No major impacts on historical sites in the area were predicted, with the exception of some very minor disruption of cemeteries through construction of sewer lines. A secondary consequence of this assessment procedure was recommendation for registration of certain historical sites on the National Register.

With respect to the physical environment in the study area, the major impact of the proposed plan was predicted to be a significant improvement in overall water quality. However, some detrimental impacts were also identified, including a minor long-term detrimental impact on air quality in the study region if a sludge incinerator were operated as part of the proposed facilities. Additionally, a detrimental impact on land resources in the study area was predicted for the disposal of wastewater treatment sludge from the proposed plants. The construction activities associated with implementation of the proposed plan were considered to have potential short-term effects of relatively minor importance. The most significant detrimental short-term impact was predicted to occur in portions of the study area where the bedrock topography was very steep and irregular and the surficial soil layers were shallow and highly subject to erosion. Blasting activities were considered to be necessary to excavate for sewer lines in the irregular bedrock topography, and significant erosion of the shallow soil layers in these areas was also predicted to be a likely consequence of construction activities. A possible detrimental impact was also predicted with respect to groundwater resources in the area. This impact was considered to be of a more permanent nature than the impacts associated with construction activities. It was felt that construction of sewer lines through groundwater recharge areas in the study region would lead to possible degradation of groundwater quality through leakage of wastewaters from the sewer lines into the groundwater aquifers.

The effects of the proposed management plan on surface water resources, wet lands, floodplain areas, prime agricultural lands, areas of high population density, and heavily used corridors in the study area were evaluated in a manner similar to that described for the groundwater resources. In all of these investigations, major attention was directed toward quantitative evaluation of impacts from the proposed plan; with respect to the alternative plans, the estimates of impact tended to be of a more qualitative nature. A quantitative estimate of overall impact for each of the alternatives was obtained through application of the rating system previously described.

6-4.8 Summary

The impact assessment for this case was completed with a description of the predicted relationship between local, short-term uses of the environment and the maintenance and enhancement of long-term productivity which would be caused by the implementation of the proposed water quality management plan. Since the plan was designed primarily to enhance long-term water quality, the short-term disruption of the environment necessary for construction of the proposed facilities was considered to be necessary and justifiable in view of the long-term benefits to be obtained. The irreversible and irretrievable commitments of resources which would be involved in the proposed plan were investigated. These commitments included allocations of land, buildings, equipment, chemicals, manpower, and energy involved in the construction and operation of the proposed treatment facilities and collection systems. However, it was considered that most of the land committed to this system would be affected only temporarily, in terms of long-term impact, since many of the facilities could be altered or removed in the future.

This case serves to illustrate well the possibilities of simplifying the assessment of environmental impact through the completion of many of the tasks associated with such assessment during comprehensive environmental planning programs. This assessment was completed as the culmination of a comprehensive water resources management plan. It served as a summary statement to the public, representing the results of years of preparation by the planning agency of the municipal area under study, the consulting engineers hired by the planning agency, and the citizens' advisory committee who had assisted the consultants in developing the final management plan. Many other opportunities exist, for other types of projects, to perform environmental assessments on proposed projects as the culmination of the total planning process for such projects. If public involvement is achieved in the beginning of the total planning process, and maintained continuously during the duration of the process, the completion of the environmental assessment study and the preparation of an impact statement describing the predicted effects of the proposed plan will be a most effective means of summarizing the total planning activity and making it understandable to the general public.

6-5. CASE FOUR—DAM CONSTRUCTION

This case history is concerned with a project on a major river in the Midwest. The proposed action in this case was the construction of a dam and locks across the river to improve navigational characteristics of the stream. The new dam and locks were intended to replace three smaller dams and locks located upstream of the new facility. The pool created by the new dam was to be more than 100 miles long and would extend significantly upstream along tributaries to the major river. The new dam would create a much deeper pool at the downstream end of the impoundment and would facilitate navigation by decreasing the time required for commercial traffic to pass through the stretch of stream under consideration. In the existing dams, the lock structures were relatively short, and it was necessary for commercial traffic consisting of long strings of barges to be broken into small units and transferred piecemeal through the small locks. A considerable amount of time was required for such operations, and the replacement of three small locks with one lock large enough to accommodate long strings of barges was considered to be a great benefit to commercial navigation on the stream in question. At the time of the passage of NEPA, the lock and dam structures had been under construction for approximately 9 years; the construction project altogether took 13 years for completion.

At the time of the environmental assessment, the lock structures had been completed and the dam was nearing completion. The new structure would increase the surface area of the impounded stream by about 5%, but the volume of the new deeper pool would be approximately 70% greater than that of the existing pool. At the time of the assessment, all lands to be inundated by the new pool had been purchased, or easements had been obtained for inundation.

6-5.1 Environmental Setting Without the Project

The environmental setting of the proposed project was a major river valley, created principally as a result of drainage of melt waters from glaciers, approximately 15,000 to 20,000 years ago. The major geologic structures of the area include gently dipping sedimentary rock layers of limestone, sandstone, and shale, across which the river valley meanders. Since the bedrock layers are inclined, erosion has exposed a time sequence of materials from one end of the impoundment to the other. Within the river valley itself, the bedrock valley has been cut to a depth considerably below that of the present river valley. In many places, the bedrock is as much as 100 feet below the present valley floor. The valley above the bedrock has been filled with sand and gravel, material principally of glacial age. Older terrace deposits of sand and gravel are found in areas along the river valley. In places, the river meanders close to bedrock cliffs, but such exposures of bedrock are relatively rare. In some places, the uplands surrounding the valley are covered with residual soils and windblown silts. Included in the study area was an exposure of very rare fossils at the extreme upstream end of the impoundment where the stream had created a waterfall.

The river valley has been for many, many years a focus of culture and occupation. Archaeologists have found numerous evidences of temporary and permanent occupation sites along the terraces and uplands in the river valley. The abundant fauna and floral resources of the stream valley were extremely attractive to prehistoric peoples. In addition to the prehistoric utilization of the river valley, intensive utilization of the resources in the impoundment area began three centuries before the time of the environmental assessment, with the coming of white explorers into the area. Because of the ease of transport of materials along the river valley, the valley was the site of many early settlements. The history of the river valley is particularly rich in episodes of significance during Colonial times and during the westward expansion which followed the Revolutionary War. Additionally, the river valley was a focus for much conflict during the Civil War. The population of the river valley experienced a slight decrease immediately following the Civil War, and has steadily increased up to the present time. Increasing utilization of the river as a means of transportation for bulk commodities has led to increased development of the region.

The present inhabitants of the river valley have lived there most of their lives and have owned property for more than one generation. The predominant life style within the river valley is rural. Most of the people surveyed during the socio-economic assessment carried out on this project were more than forty years old, and nearly two-thirds of them were married. Most of the heads of households interviewed were married, male farmers. Near the upstream end of the impoundment, a major city was located, and the socio-economic environment was considerably different from that of most of the region affected by the dam. For example, in one rural county located in the middle portion of the impoundment, more than one-fourth of the population had lived in one house for more than 20 years; in contrast, in the municipal area located at the upstream end of the impoundment, only about one-twentieth of the population had lived in one house for more than 20 years. Only in the industrialized counties near the upstream end of the impoundment did the median family income substantially exceed a poverty level of $3,000 per year.

The existence of several low-level locks and dams in the proposed impoundment area allowed management of the stream, to store flood waters to a minor degree, but principally to maintain navigational capabilities. Because the dams were essentially low-rise structures, the locks generally were in operation during periods of low flow, but when high water levels persisted, the wickets at the dams were lowered, and commercial traffic could bypass the lock structures. For the most part, the locks were utilized during summer and fall, and open navigation conditions generally were experienced in late winter and early spring and summer. Generally, data on travel times of water through the pools showed that the shallow dams had little effect on the river's flow pattern. The shallow dams had some beneficial effects on the quality of the water in the impoundment area, however. In general, the quality of water in the impoundment area had been consistently degraded as a result of wastewater discharges from the urban industrial areas near the headwaters of the impoundment. Domestic wastewaters and industrial wastes from various facilities in the upper portion of the impoundment area were discharged into the river in significant amounts. These discharges caused the stream quality to deteriorate over the years. For example, at the time of the assessment study, the dissolved oxygen content in the stream represented saturation values of only 35 to 65%. Additionally, more intensive industrial utilization of river bank sites was contemplated at the time of the environmental assessment. The existence of the significant pollutant concentrations and the depletion of dissolved oxygen in the stream had had a very detrimental effect on wildlife in the stream itself and on the banks, at the time of the environmental assessment.

The flora and fauna of the impoundment area were fairly representative of a transitional area between Appalachian environments and Midwest Plains environments. The flora in the study area is quite diverse and is composed of more than 100 different families of plants. Although there is not a great abundance of species of rooted plants in the river valley itself, there is a great abundance and diversity of algae. In assessment studies, more than 150 species of algae were identified in the impoundment area. The existence of abundant algae has been attributed to the high levels of pollutant nutrients in the waters below the urbanized area at the upstream end of the impoundment. In contrast to the algae, rooted plants are not favored because of the fluctuating water levels in the stream and because of frequent flooding. Additionally, the stream carries a rather heavy silt burden, and the resultant turbidity is unfavorable to rooted plants. The banks of the stream, however, are rather densely covered with many varieties of riparian vegetation, including cottonwoods, sycamores, elms, elders, willows, and some alders.

The diversity of plant food sources in the stream valley has produced a diversity of animals in the area, also. With respect to the aquatic environment, the pollutants released by the industrial facilities and the large urban areas at the upstream portion of the impoundment area have had detrimental effects on bottom organisms in the stream. However, over 54 species of zooplankton were identified during the assessment studies. The most common zooplankters were rotifers. In the impoundment area, macroinvertebrate fauna were sparse, particularly in the polluted area of the stream. The fish in the river represent a number of diverse species and are relatively abundant. Over 130 species of fishes were identified in the proposed impoundment area. Additionally, over 35 species of amphibians and over 40 species of reptiles were identified as inhabitants of the proposed impoundment. Likewise, the birds in the study area were abundant and diverse, including over 296 species. No rare or locally indigenous species of mammalian fauna which did not exist elsewhere were found in the impoundment area. Most of the indigenous larger animals that once occupied the impoundment area have vanished since the intensive usage of the area during the last two centuries by white settlers.

This, then, was the environmental setting of the project, prior to the beginning of construction of the proposed dam. At the time of the environmental assessment, the dam had been virtually completed, and much of the guesswork had been taken out of the assessment, since many of the impacts of the construction were evident.

6-5.2   Impacts of the Proposed Action

The assessment of the virtually complete locks and dam structure and the impounded river above the dam showed that the environmental effects of the construction had been significant. Changes in erosion and sedimentation patterns already had been created within the river channel itself. In addition, significant areas of floodplain farmlands had been inundated; these flood plains are among the most fertile agricultural areas within the region. Also, the river banks were being eroded at an accelerated rate at elevations above the prior shorelines, with some severe localized erosion and caving of the river bank immediately downstream from the completed dam on one side of the river. The erosion was caused by the change in elevation of the contact between water and soil, and a new equilibrium was being established between the soil and the water. However, some beneficial effects were created with respect to the soil and rock of the river banks, by the change in elevation within the impoundment. Groundwater levels throughout the alluvium in the valley along the impoundment length rose and power requirements for pumping of groundwater were reduced. Additionally, the overall stability of the river banks was increased. Massive slope failures of the river bank were found to occur most frequently after flooding, when the saturated river banks were subjected to the combined forces of seepage and gravity. Since the river was to be maintained at a higher level, as a result of the dam construction, the drop of water elevation after flooding would not be as great as it had been formerly, and instability problems in the river bank would be decreased. Additionally, the impact of the constructed facility on any of the unique geological materials within the impoundment, including the rare fossils mentioned previously, was found to be minimal because of an insignificant rise in water level at the fossil exposure. The increase in elevation of the river in the impoundment area, however, was found to have some dire effects on some other aspects of the environment.

Over 700 occupation sights were found during the archaelogical exploration of the impoundment shorelines. Since the water level had been raised in the lower portion of the pool area at the time of the assessment study, it was felt that many archaeological sites in that area were not discovered. The change in water level, leading to the bank erosion mentioned previously, had an extremely detrimental effect on the archaelogical sites located on the river banks and immediately above the river banks. Also, the inundation of low-lying areas effectively eliminated any archaeological sites in those areas, since they would be lost to exploration and investigation, by submergence. In contrast to the prehistoric sites found to be adversely affected by the impoundment, no significant historic sites were found to be affected.

With respect to socio-economic effects, major impacts centered around the interactions among the landowners and the personnel of the agency which had initiated and constructed the dam and locks. More than 1,000 landowners were involved in either purchase agreements or easements, in connection with this project. More than 20% of the population surveyed during the assessment indicated that their first knowledge of the project came through contact with a real estate purchasing agent who arrived at their homes to discuss purchase of property. Less than 2% of the homeowners involved in the project had heard about the contemplated action from a formal source. Additionally, both the landowners and the real estate agents charged with purchasing land were somewhat misinformed about the procedures involved in land acquisition and the overall purpose and effects of the proposed project. Although more than one-third of the population contacted in the assessment thought they should have had some input in the planning and development process for this project, no provision had been made for citizen involvement. As a consequence, a hostile and bitter attitude toward the initiating agency had developed among the landowners. Some of the persons living within the impoundment area were directly affected through changes in the characteristics of the river basin. For some people, favorite fishing spots had disappeared; for others, relocation to new housing sites was required. These two effects represent extremes in socio-economic impact. Many of the people who were relocated because of the project had intended to live on their purchased property and to retire there. Thus, the social implications of relocation were severe for the small number of families who were relocated. Finally, many persons in the impoundment area indicated strong opinions concerning a change in the aesthetic characteristics of the area. Some of the persons interviewed considered that the beauty of the pool area was increased by the dam, while others stated, just as vehemently, that the beauty of the area had been severely marred.

Although the social aspects of the project were rather mixed, the local economy was provided a stimulus through the construction project, with an influx of more than $20,000,000 in salaries and wages. However, the replacement of three low-rise dams with one high-rise dam was estimated to produce an annual loss of over $500,000 in direct income in the impoundment area. Most of the land purchased in the area was valuable farmland, but some of the acreage had been idle prior to purchase. Consequently, it is difficult to assess the overall losses and gains associated with purchase of land for the impoundment. The benefits of improvements in navigation capabilities on the stream have been estimated at more than $5,000,000 each year, on the basis of savings to producers, shippers, and consumers of goods transported on the river.

With respect to the quality of the aquatic environment, the reduction of velocity and the increased depth in the pool created by the new dam was considered to have an adverse effect on the quality of the stream water. In contrast, the development of phytoplankton blooms in the larger and deeper pool area was considered to produce a beneficial effect on the stream. The increased area of the pool and the greater width of large open areas of water were estimated to be beneficial in assisting aeration through greater turbulence and wave action. In addition to improving aeration in the stream, the phytoplankton blooms were assessed as beneficial from the point of view of removing nitrogen and phosphorus from the stream and making them available to higher-order organisms in the river. Additionally, construction of the dam was estimated to have a beneficial effect on turbidity since the lower velocity in the deeper pool was expected to clear up the water considerably. Of course, the removal of the three low-rise dams, with their provisions of overflow during high-water periods, decreased the potential for aeration of the river water in the impoundment area and was judged to have a detrimental effect on the impoundment water quality.

A major feature of the assessment study for this project was rather dramatically presented early in the study effort when the contrasting effects of the new dam and the low-rise dams on water quality were evaluated. The study effort then centered on an evaluation of the relative benefits and detrimental effects of the three low dams versus the high dam. A computerized model was utilized to simulate stream flow under the conditions existing prior to construction and for conditions after construction of the dam. The model was calibrated with data obtained prior to the construction of the dam. The overall results of the modeling showed that at very low flows in the river, a severe deficit would develop in dissolved oxygen in the lower reaches in the pool. This result was postulated on the absence of an active photosynthetic output by organisms in the river. Thus, one of the results of this assessment effort was to focus further investigation on the possibility for organic photosynthesis in the impoundment reaches. The model study showed that for higher volumes of flow, very little difference would exist in water quality in the impoundment as compared to the preconstruction environment.

With respect to the terrestrial flora and fauna in the impoundment region, the effects of the change in water levels was considered to be minimal for the upland vegetation; most of the species involved are adapted for periodic flooding and are likely to become established on higher ground prior to their erosion and displacement from lower levels. At the time of the survey, a significant number of riparian trees had been removed from the bank, as a result of undercutting through erosion. These trees were up to 36 inches in diameter and constituted serious hazards to small boat navigation, especially at night. Additionally, the removal of these large trees from the river banks was considered to have a detrimental effect on the birds and animals which utilized the trees as nesting sites and den sites. In contrast to the effects on stream-side trees, the submersed species of the aquatic plants were considered to be affected favorably by the creation of the new impoundment, since more extensive embayments would be created at the mouths of tributary streams. Additionally, some marshes and sloughs were enlarged through construction of the dam and impoundment of the new pool.

With respect to organisms, the increased area and depth of the impoundment at its lower end, with an increased retention time for water, was postulated as a favorable factor for the growth of phytoplankton. This increase in algae should have an overall beneficial effect in increasing faunal productivity in the river. Also, the removal of the low-rise dams and the impoundment of a deeper lake-like body of water should increase the spawning areas for some species of fish which are common to the area. The increase in phytoplankton will bring about an increase in zooplankton, which in turn will provide food for bottom fauna and fishes. An important benefit that could be realized from the construction of the new dam is the possibility for exploitation of the increased fish population in the stream, through commercial fisheries.

In contrast to the fish and other aquatic organisms in the stream itself, the amphibians, reptiles, birds, and mammals that inhabit the regions around the impoundment are not likely to be affected significantly by construction of the dam.

With respect to the cultural aspects of the environment not previously mentioned, the assessment team considered that creation of the new dam and pool would enhance the aesthetic qualities of the portion of river affected. Although some traditional recreational areas would be lost, the establishment of a deeper, wider, and more stable pool should enhance the beauty of the river environment and should lead to greater recreational usage of the river. The creation of wider and longer expanses of water, and the decrease in turbidity expected in the new impoundment certainly should add to the beauty of the stream.

6-5.3   Other NEPA Aspects

The other aspects of environmental assessment required by NEPA include identification of unavoidable effects and similar considerations. These effects are mentioned in the following paragraphs.

Unavoidable Adverse Effects on the Environment. Included in the unavoidable adverse impacts of this particular facility were the inundation of valuable land, changes in sedimentation regime of the river, and erosion within the stream channel and along the river banks. Additionally, erosion of the river bank has caused undercutting and collapse, as well as actual removal, of valuable archaeological materials. The erosion of such materials destroys their significance in the reconstruction of early cultures. The process of acquiring land which would be inundated or for which a flowage easement would be required apparently was ill-managed. Thus, an adverse social effect was caused through the generation of hostility and anxiety in the minds of landowners who were contacted for purchase or lease of land. This lack of public involvement, with its consequent public hostility to the project, could have been mitigated, but it was not. The principal unavoidable adverse economic consequences of the project included the loss of inundated land and the loss of income as jobs were eliminated with the removal of low-rise locks and dams within the impoundment area. Finally, the effect of the impoundment on the water quality of the stream was uncertain, but a significant possibility existed for a deterioration of water quality as a result of increase in retention time of water in the impoundment, and deepening of the water body in the lower reaches of the pool during low flows.

Alternatives to the Proposed Action. Since the project was virtually complete at the time of the environmental assessment, there was very little consideration of alternatives to the proposed action. The only alternatives which could have been implemented were operation of the river system as if the dam were not in existence, or operation of the facility to manage flood waters as well as to improve navigation. In this instance, it was deemed to be inappropriate to consider operating the river system as if the dam and locks were nonexistent, since no benefits could accrue from such action. On the other hand, the initiating agency has expressed a willingness to operate the facility to increase flood water management capabilities, as well as to maintain navigational capabilities in the river system. Although no other alternatives were available in this case, the case has served as a valuable study, in that many predictive methods and techniques were utilized to forecast detrimental effects, based upon conditions known to have existed prior to construction. These predictions were then compared with actual effects and the prediction methods were refined. The benefit gained from such validation of prediction techniques was very important in this case, since two other major dams were planned or were under construction for the same river system. Experience in assessing the impacts of the facility which was virtually complete was of great value in later assessment of impacts for structures in earlier stages of construction.

Relationship between Local Short-Term Uses of Man's Environment and the Maintenance and Enhancement of Long-Term Productivity. With respect to the physical system, the short-term effects of this project were virtually minimal. In contrast, long-term effects were judged to be quite significant. The raising of the water level in the stream and the creation of a lake over 100 miles long should provide better capability for management of flood waters and will raise the groundwater levels in the lower reaches of the pool, thus enhancing water supplies in that region. The stable pool should enhance the value of shoreline property. In the hydrologic aspects of the project, the overall increase in stream stability has been judged to be a benefit greatly outweighing any negative effects on the water resources of the area. The archaeological effects mentioned previously are considered to be of long-term significance because the archaeological materials are permanently lost when the sites are eroded. The major adverse social impacts, arising from lack of citizen involvement, were created over a relatively short time but have been judged to be of long-lasting effect. The relocation of persons forced to move has had a permanent effect upon their attitudes toward the agency initiating and completing the project. Throughout the study region, the inhabitants of the area have expressed a general lack of trust of the agency involved and expressed feelings of having been misguided and misinformed about the project. With respect to economic conditions, the localized boost to the economy, as a result of construction operations, will be accompanied by a rapid loss of jobs for workers supplying services to construction force personnel. A cycle of rapid expansion during early stages of construction will be accompanied by an equally rapid depression phase after completion of the project. However, the improvements to navigation created by the project may have significant long-term economic benefits, through the enducement of additional investment in the impoundment area. Additionally, the project may beneficially affect regional economic integration. The initial investment in the modernizing of the dam and locks system may produce a multiplier effect in the area. With respect to the biota of the area, the increased retention time of water in the impoundment, with the greater depth and width of the new pool, should increase long-term productivity. There was a good indication during the assessment study that the pool may be capable of sustaining commercial fisheries in the future. The general long-term effect of the development on the plants and animals of the region should be beneficial.

Irreplaceable or Irretrievable Commitments of Resources. The establishment of the dam and lock structure permanently raised the river level in the impoundment. This created a permanent inundation of almost two square miles of bottomland and adjacent area. This inundation has removed certain fertile land from use, and thus, represents a commitment of that resource which is irretrievable during the life of the facility. Any mineral resources found within the area of inundation also must be considered to be lost during the life of the structure. The commitment of certain quantities of construction materials during the building of the structure can be considered to be an irretrievable commitment of those resources, but the significance of this allocation of resources is rather slight. No irretrievable or irreplaceable commitments of fauna or flora were created during the construction of this facility. An irretrievable loss of archaeological resources was felt through inundation and destruction of archaeological sites in the bottomland and on the banks affected by this impoundment.

6-5.4 Summary

In this case history, the assessment of a virtually complete facility has been described. This case history can show the range of effects possible in the impoundment of a large stream. Additionally, this case serves to illustrate the benefits to be gained from assessment and evaluation of a facility which is virtually completed. The major benefit gained in this instance was a validation of assessment techniques, which then could be used to investigate other structures of a similar nature planned for subsequent construction in the same area. Additionally, this case showed very clearly the extremely detrimental consequences associated with neglect of the public during planning and implementation studies for a major project. Although only a few persons were directly affected detrimentally by the project, a general feeling of mistrust and hostility toward the

initiating agency was evident in the inhabitants of the impoundment area. This unfortunate situation could have been eliminated
with adequate efforts to inform and involve the public at an early
stage of planning of this project. It must be remembered, in this
context, that this particular project had been under construction
for approximately 8 years at the time of passage of the National
Environmental Policy Act. These detrimental effects from inadequate public involvement should serve as good indicators for the
planning of assessment studies and for the overall planning of
activities by management agencies. It is always important to learn
from mistakes. In this case, a major mistake was made in inadequately involving the public in the contemplated action. This case
shows a vivid contrast with the comprehensive public involvement
program described in Case 3.

6-6. CASE FIVE—OPEN BURNING

This case is presented to show how the basic methodology and philosophy of impact assessment can be utilized for the evaluation and
analysis of problems which are smaller in magnitude or less significant in effects than some of the projects mentioned in the previously cited case histories. This case was developed from a study
undertaken because of the restriction of construction activities, in
a particular county within a state, by the actions of a local governmental regulatory agency. No federal funds were involved, nor
were federal jurisdictions involved. However, the basic philosophy
of impact assessment, as outlined in this book, was applicable in
this case and should be considered applicable in similar cases.

6-6.1 General

The regulatory body charged with controlling air pollution in a
county in an eastern state adopted control measures, including a
prohibition against open burning of refuse accumulated on construction projects. Included in this refuse were the trees and brush
and similar materials cleared from the natural site. As a result of
protest by local contractors and developers, a period of one year
was granted before full implementation of this prohibition of open

burning, in order to allow the affected individuals to develop alternative disposal methods. Several of the methods suggested by members of the regulatory agency appeared to be extremely costly from the viewpoint of the contractors involved. At this point, it was deemed appropriate to conduct a study of alternative disposal methods for site clearing wastes, utilizing the basic philosophy of impact assessment as outlined herein.

The first major item of effort in the study was to establish the environmental setting in which the contemplated change in operations on construction sites would take place. For the purpose of this particular study, the environmental description consisted principally in establishing the characteristics and amounts, as well as the geographic distribution of site clearing activities (and site clearing wastes) in the county under consideration. Additionally, the description of the current status in the study area included a description of the current methods of disposal for clearing wastes. This description included an analysis of the potential for air pollution, water pollution, land misuse, noise pollution, and aesthetic insult associated with currently utilized disposal practices. This study effort constituted a parallel effort to the description of the existing environment without a project, as included in environmental impact statement studies. The next item of study was a definition and description of viable alternative methods of waste disposal.

6-6.2 Alternatives

The alternatives considered for the study of waste disposal included the following methods:

1. A continuation of uncontrolled open burning as practiced in the past;
2. Use of on-site incinerators and combustion methods, significantly better, from a pollution control viewpoint, than the uncontrolled open burning currently in use, with burial of any residue remaining after incineration on the construction site;
3. The collection, appropriate preparation, and transfer of waste to a series of satellite incinerators located at appropriate points within the area of site clearing activity;

4. The collection, appropriate preparation, and transfer of wastes to a centrally located incinerator facility; and
5. The collection, appropriate preparation, and transfer of wastes to available sanitary landfill disposal sites.

These alternative methods were considered as viable modes of operation and were analyzed in the study effort. It had been suggested that the most advantageous disposal of site clearing wastes would be through a recycling of the wood materials in pulping operations or in similar activities. A significant effort was made in connection with determining the feasibility of such recycling. It was established very early in the study that 1) no processor for site clearing wastes in the form of mixed trees and brush was operating in the vicinity of the county under consideration, 2) no interest in such operation was expressed by any of the individuals contacted during this phase of the study who were engaged in similar salvage operations on other types of waste materials, and 3) the nearest market area of site clearing wastes in the county under study was approximately 200 miles away, and the cost of transport of clearing wastes to such a remote facility was very significant. For these reasons, the recycling of the wood wastes was not considered to be a viable alternative, and the remainder of the study was devoted to consideration of the alternatives listed above.

6-6.3 Evaluation of Alternatives

On the basis of existing information, it was obvious that over 1,000 acres of densely wooded land was being cleared each year in the county under study. It was estimated that approximately 75 tons of trees would be removed from each acre of forest land and, additionally, approximately four tons of brush, leaves, and litter would be accumulated on each acre. Finally, approximately two tons of small trees, from one to eight inches in diameter, would require clearing from each acre of developed land. Thus, approximately 80 tons of clearing wastes would require disposal from each acre of developed land. For the entire county area, clearing requirements would amount to about 80 thousand tons annually. The various possible ways to dispose of these wastes are described in the following paragraphs.

Open Burning. Continued utilization of uncontrolled open burning of site clearing wastes was found to present serious impact, from the point of view of physical and chemical parameters, socio-economic parameters, and aesthetic parameters of environmental quality. The principal impact, in the physical and chemical spheres, of the continued use of open burning would have been a significant production of air pollutants, including, principally, carbon monoxide, unburned hydrocarbons, and significant amounts of particulate matter. The production of such air pollutants would have had significant socio-economic effects in the county under study. A serious political inequality would have been associated with the open burning of site wastes by one segment of the community (home builders and contractors) and the prevention of any open burning of wastes on the part of residents of the county (residents are even forbidden to burn leaves in the fall). Additionally, continued open burning of clearing wastes was considered to be detrimental, in that it could produce physiological damage in individual citizens, from the generation of air pollutants; on this basis, continued open burning was considered to be a safety hazard to community health. Finally, the odor and visual effluents produced by uncontrolled open burning were considered to have a very negative impact, from the point of view of aesthetics; this impact would certainly disrupt the composition and mood of a residential area adjacent to the site being cleared. In contrast to these detrimental impacts, the continued usage of open burning would represent the most economical and convenient method of disposal of such site clearing wastes.

Controlled On-Site Combustion. The second alternative considered was the disposal of site clearing wastes through carefully controlled combustion operations carried out on site. The controlled combustion was to be achieved through the use of devices such as air curtain destructors. The physical and chemical impacts of such practices was considered to be negligible with respect to the site being cleared and the area surrounding the site. The impacts essentially were assessed as minimal, because of the relatively insignificant production of air pollutants through carefully controlled combustion. Only in a very small area (generally a pit) where the controlled combustion operations were carried out would any detrimental impacts be felt. At the particular spot where the combustion would occur, a very minor impact would result from the production of a baked area in the soil and the generation of a very slight amount of residual matter from the combustion. Only negligible amounts of air pollutants would be generated with proper operation of combustion units such as air curtain destructors. The socioeconomic impacts of controlled on-site combustion were considered to be relatively minor, in comparison to the effects of continued uncontrolled open burning. A significant impact was predicted in the economic sphere, with respect to the increases in costs for the cleared sites to be purchased by the public. The increased capital costs for controlled combustion devices and the installation costs for required pits, labor, and appurtenances obviously would be passed on by the contractors and site developers to the public. Because only insignificant amounts of air pollutants would be generated through proper operations of on-site combustors, the physiological health problems and safety problems associated with controlled burning would be minimal. Because of the combustion within a pit, and because of the elimination of smoke and other air pollutants, through the use of controlled combustion operations on site, the aesthetic impact of controlled combustion disposal of clearing wastes was considered to be minimal.

Combustion in Satellite Incinerators. One of the alternatives considered in this study was the use of a series of small satellite incinerators, constructed at appropriate spots throughout the area of clearing. The disposal of clearing wastes in this way was considered to be preferable to the continued utilization of uncontrolled open burning. However, it was not considered to be as advantageous as carefully controlled combustion on the construction site. The impacts of disposal in satellite incinerators, with respect to the physical and chemical parameters of environmental quality, were assessed as more extensive than would be the impact of controlled combustion on site. Land would be required for the physical plants of the satellite incinerators. It was considered to be difficult to obtain adequate zoning for such utilization of land. This was considered to be an associated socio-economic difficulty related to the use of satellite incinerators. Additionally, because it would be necessary to transport cleared wastes from the construction sites to the satellite incinerators, it was necessary to assess the environmental impacts of the transfer operation. Although the operation of a well-controlled incinerator would produce only slight amounts of air pollutants, significant amounts of air pollutants, such as carbon monoxide, unburned hydrocarbons, nitrogen oxides, and photochemical oxidants, would be produced through the operation of internal combustion engines in the vehicles used to transport the wastes from the construction sites to the satellite incinerators. Additionally, significant amounts of noise pollution would be generated on the construction sites through the necessary use of chain saws required for reducing large trees to manageable pieces for shipment. It was suggested that a shredder could be used to reduce the sizes of the pieces of material so that pay loads in transfer vehicles could be increased. If such a grinder or chipper were used, very significant levels of noise were predicted for the construction sites. Additionally, the truck traffic generated through transfer of wastes from the construction sites to the satellite incinerators would create a significant amount of noise along the streets and roadways travelled by the transfer trucks. The socio-economic effects of this alternative method of disposal would include somewhat higher costs for land clearing, and these costs would be passed on to the buyers of the cleared land. Additionally, sociological impacts would be felt through the routing of truck traffic through neighborhoods of residential character to satellite incinerators. Significant impacts on individual well-being also were predicted. The generation of air pollutants and noise from transfer vehicles was considered to be detrimental to the health of the persons along the transfer routes. Additionally, a significant hazard would be created through the increase in truck traffic along the transfer routes. Other social effects included possible adverse community reactions to the location of satellite incinerators in semi-developed areas. The mood and organization of a particular community could be seriously disrupted by the construction of a satellite incinerator and the generation of truck traffic to that facility from surrounding land-clearing sites. The aesthetic impacts of the use of satellite incinerators included the possible incompatibility of the satellite incinerator plants with the present development and existing structures in the areas in which they would be located. Additionally, the impact of satellite incinerator construction and the generation of truck traffic in existing residential areas would have been very negative from an aesthetic viewpoint.

Incineration at a Central Location. Essentially, incineration at a central location would involve many of the same impacts as felt through the use of satellite incinerators. However, because of the greater probability of long-distance hauling of site clearing wastes from remote suburban areas to a central municipal incinerator, it was assumed that clearing wastes would be reduced in size through grinding or chipping prior to transfer. The grinding and chipping of the wastes would certainly increase payloads in transfer vehicles, but, as mentioned previously, would create significant levels of noise on site. As in the case of combustion in satellite incinerators, only insignificant amounts of air pollutants would be produced during the combustion operation, if it is assumed that the centralized incinerator is a well controlled facility. However, very significant amounts of air pollutants would be generated from the engines of transfer vehicles required in this scheme. Additionally, the charging of the municipal incinerator with very significant quantities of clearing wastes was found to be exceedingly detrimental, in that it would conflict with the main function of the municipal incinerator in the area under study, that is, the disposal of the municipal domestic solid wastes. In fact, it was discovered that in the particular case under investigation, the operators of the municipal incinerator would not accept site clearing wastes for disposal. Significant impacts in the socio-economic sphere also were predicted through use of the municipal incinerator for the disposal of site clearing wastes. It was predicted that an undesirable social reaction would be provoked by the interference with solid waste disposal capabilities at the municipal incinerator. Since much of the land clearing was being done in essentially rural or undeveloped areas of the county under study, the residents of the urban area who rely on the incinerator for solid waste disposal would, in all likelihood, react very unfavorably to the overloading of this facility with wastes cleared from suburban construction sites. In evaluating this problem, it was apparent that the volume of clearing

wastes produced in the county under study each year would severely tax the capabilities of the existing municipal incinerator. As
in the case of the satellite incinerators mentioned above, safety
hazards would be created through existing residential and densely
populated urban areas. Aesthetically, the major effect of disposal
at a central incinerator would consist of the undesirable noise associated with sawing and shredding of wastes on site and the noise
produced by transfer truck traffic. Additionally, some general undesirable aesthetic impacts would be felt through the transport of
waste materials from all points of the suburban fringe in the county
to a central urban location.

Sanitary Landfilling of Site Clearing Wastes. The last alternative
considered in this evaluation was the collection of wastes on site;
the preparation of wastes for transfer, through sawing and shredding; and the subsequent transport of waste materials to sanitary
landfills for ultimate disposal. The physical and chemical impacts
of such disposal could include possible detrimental effects on
water, land, air, and noise. From a practical standpoint, disposal of
clearing wastes in landfills would be very difficult because of the
low densities achieved in compaction of such materials. Also,
these wastes would be highly biodegradable. The production and
migration of leachate from a landfill consisting of organic clearing
wastes would be highly probable, and such migration of leachate
would have significant undesirable effects upon groundwater
resources near the fills. Leachate production would be likely to
occur in landfilling of site wastes, even without the infiltration of
external water, since water would be produced during the decomposition of brush and trees cleared from the construction site.
Additionally, landfilling of such waste materials would rapidly
exhaust the available land spaces reserved, at the time of the assessment, for the disposal of municipal solid waste. The landfills
used for disposal of site clearing wastes would be rendered unfit for
future use for a considerable period of time because of the detrimental material decomposition which would occur in the highly
degradable refuse. This decomposition would produce undesirable
gaseous by-products and settlement of the land surface over the fill.
As in the two previously mentioned methods which involved trans-

fer of materials, in this method, a significant amount of air pollutants would be generated during the truck transfer of collected wastes to remote disposal sites. Additionally, significant amounts of noise pollution would be produced by the use of saws and shredders on the construction sites, by the truck traffic involved in transport of collected wastes, and by the operation of landfilling equipment at the ultimate disposal sites. Finally, it was found that a severe physical limitation on the availability of land for such disposal existed in the county under study at the time of the assessment. Additional undesirable environmental impacts of landfilling of clearing wastes included the possible generation and spreading of pest species such as termites, which would be supported in timber and brush wastes.

The socio-economic effects of landfill disposal would include the increase in purchase price for new home sites associated with the increased costs of landfilling and transport to landfills. Additionally, since it was predicted that landfill sites within the county under investigation would soon be exhausted, long-term utilization of landfills would require travel across county boundaries, and the possible political ramifications of such transport were considered to be highly undesirable. Additionally, increased safety hazards were predicted through the generation of truck traffic for refuse disposal. Increased noise hazards would be created, as would possible groundwater pollution hazards near landfills. The aesthetic impacts of landfill disposal would include some change in the topography of the filled area, whether beneficial or detrimental. The change in quality of the topography and landscape in the fill area would depend upon the skill with which the filling operation was conducted. Undesirable aesthetic impacts would be created from the generation of noise during the transport and filling operation.

6-6.4 Further Studies

After completion of the evaluation of alternatives as outlined above, it became apparent that on site combustion under very carefully controlled conditions could represent the most environmentally and economically sound method of waste disposal. To

(a)

(b)

Fig. 6-3   An air-curtain destructor used for controlled open burning. This device offers a viable alternative to uncontrolled fires (*courtesy* Mr. K. A. Barker).

267

fully evaluate this alternative, a series of tests were conducted in the county under study, wherein site clearing wastes were reduced in conventional, uncontrolled open burning operations, and through the use of an air curtain destructor. The quality of the effects on the surrounding environment were measured through constant measurement of Ringlemann smoke density numbers as indicators of the production of air pollutants during combustion. Ringlemann numbers were used at the behest of the regulatory agency. Observation of uncontrolled open burning operations indicated that Ringlemann smoke density numbers of "three" or "four" were obtained approximately 85% of the time during which the burning took place. In contrast, utilization of a pipe blower, fabricated by a local contractor, showed Ringlemann readings of "zero" 4% of the time, "one" 55% of the time, "two" 39% of the time, and "three" 2% of the time. Finally, use of an air curtain destructor showed Ringlemann readings of "zero" (no visible smoke) 97% of the time. Ringlemann readings of "one" were recorded approximately 2% of the time and Ringlemann numbers equal to "two" were recorded approximately 1% of the time. Thus, utilization of an air curtain destructor was considered to produce virtually pollution-free combustion on site 97% of the time. Only during periods when fresh wastes were charged to the operating pit were Ringlemann readings greater than "one" obtained.

On the basis of the test results and the environmental assessment previously completed, it was recommended to the regulatory agency that continued disposal of site clearing wastes be carried out on site through the use of air-curtain destructor units. However, the regulatory agency chose to ignore the results of the assessment and, at the end of the one-year period of grace allowed to the developers and contractors, imposed a strict ban against all burning of site clearing wastes. As a consequence, environmentally unsound disposal of clearing wastes in landfilling operations was required.

6-6.5 Summary

In conclusion, this short case history has illustrated the utilization of the basic philosophy of environmental assessment in evaluating alternatives in a localized issue not involving federal monies. Although the decision of the regulatory agency was made in contradiction of the results of the environmental assessment, the utility of the assessment procedure, as applied to problems of local jurisdiction, should be apparent.

6-7. CASE SIX—MARINA CONSTRUCTION

The last case to be presented in this chapter represents the use of systematic environmental assessment methods for the investigation of a small project, which was not undertaken by a federal agency, but which came within the jurisdiction of a federal agency through a permit requirement. In this case, a real estate developer proposed to build a marina and housing complex on a site adjacent to a major river. Along the riverbank in this area, a number of vacation homes and weekend facilities had been occupied for a very long time by persons seeking relief from the pressures of life in a nearby large metropolitan area. Some of the housing units located along the riverbank in this area were occupied on a year-round basis. Considerable controversy arose when plans were disclosed concerning the construction of the proposed marina and housing complex. The developer of the project was informed by personnel of the local office of the U.S. Army Corps of Engineers that a permit would be required for his proposed facility, since it was to be located within the floodplain of the river and could have some influence on navigation and flooding potentials in the area. In conjunction with the "permitting" process, the federal agency officials indicated that the developer would be required to prepare an environmental assessment of the proposed marina complex. The developer engaged a number of environmental engineers as consultants to carry out such an assessment.

The consultants developed a preliminary assessment that identified several detrimental effects which could be anticipated from the proposed action. After these effects were identified, the consultants developed alternative plans to mitigate the undesirable effects of the proposed activity. These alternatives, and the null alternative, were included in the assessment procedure. The completed assessment was presented at a public meeting attended by most of the residents of the project site area. The results of this assessment investigation and the public hearing provide some valuable insights into the application of assessment methodologies for small-scale projects.

6-7.1 The Environmental Setting of the Proposed Project

The proposed marina and housing project was to be built on a site adjacent to a major river. During the glacial periods in the earth's history, when sea level dropped several hundred feet, this river had eroded the bedrock of the area to a depth of about 100 feet below the present river elevation. When sea level rose, after the ice began to melt, the river deposited granular materials including sand and gravel in the bedrock valley. During the last several thousand years of the earth's history, the river has flooded annually and deposited finer sediments over the sand and gravel. The upper 10 to 30 feet of material at the site of the proposed facility consisted of silty clays and clayey silts, which represented slackwater and floodplain deposits of the modern river.

The site of the proposed facility consisted of a depression through which an ephemeral stream was flowing. This small creek filled with water annually, during periods of flooding, when river waters backed up into its channel. However, during summer months, the stream virtually disappeared. Between the stream and its adjacent depression, and the major river, a natural levee had been created through the years by the deposit of silty material. These silty sediments were deposited by the flood waters of the major river during periods of rising water, and fine-grained materials were deposited from the water trapped behind this natural levee in the depression. The small residential community in this area occupied plots of ground along the riverbank above the high water level. A small two-lane roadway served these residences. The roadway was located on the crest of the natural levee and extended parallel to the river from a point well upstream of the project site to a point at the downstream extremity of the project site where the small, ephemeral stream emptied into the river.

The overall plan for the facility included excavation of the existing creek bed to form a channel approximately 150 feet wide and 6 feet deep. The normal water depth in the marina channel would be six feet; that is, the bottom of the channel would be six feet below the normal pool level of the river. The elevation of river waters was maintained through a series of navigation locks and dams for several hundred miles along the course of the river. A second marina channel was proposed, parallel to the course of the small stream. Thus, the overall development consisted of two parallel marina channels, approximately 150 feet wide and 6 feet deep, separated by a parcel of ground approximately 300 feet wide. The soil material removed from the marina channels was to be filled on the area between the channels and adjacent to the channels on both sides. The elevation of the ground surface would be increased so that the locations of the housing units in the marina complex would be approximately 15 to 30 feet above natural ground level.

The housing units were to consist of 400 townhouse units and 200 apartment units. Each of the units was to have a floor space of approximately 1,500 square feet. The townhouse units were to be built on land adjacent to companion docking facilities. It was anticipated that potential owners of townhouses would be attracted to the development by the convenience of adjacent docking facilities. The course of the creek was to be broadened and deepened at the point where it entered the river, so that the channel would be approximately 200 feet wide at the point where it entered the major stream.

Utilities for the housing units were to be furnished both on site and from remote sources. Domestic drinking water for the units was to be supplied through connections to the water supply network of the utility serving the nearby metropolitan area. Likewise, the electric power utility for the metropolitan region had agreed to provide electric utilities for the housing complex. Wastewater treatment facilities for the living units, and for the boats which were to be docked in the marina facility, were to be provided in a secondary wastewater treatment plant to be built at the site. This plant was to be based on the activated sludge process. Effluent from the wastewater treatment plant would be pumped into the river at an outfall point, at a location approximately 100 feet downstream from the mouth of the marina entrance channel.

6-7.2 Environmental Setting of the Project

The environmental setting of the proposed marina project was described by the consultants in terms of the geology, topography, ecology, and cultural values of the site. Additionally, land use in the project area was described. This characterization of the existing environment was considerably less comprehensive than the environmental descriptions associated with major impact statements, such as those prepared for nuclear power plants. However, the scope and depth of the description were appropriate for the small scale of this project.

Geologically, the project site was located in the floodplain of the major river described previously. This river, during the ice ages, had eroded its bedrock valley to a depth considerably below the level of the present valley. During the ice ages, sea level dropped several hundreds of feet all over the world because of the capture of water in the ice sheets. As a consequence, erosion throughout the world was accelerated, and in this area, the bedrock valley was deeply incised. When sea level rose at the conclusion of the glacial period, the river deposited a very deep filling in the valley, consisting primarily of gravels and coarse sands near the bedrock surface, with finer sands and silts closer to the present valley surface. In the last several thousand years, the river had deposited primarily fine-grained materials during periods of flooding. The characteristic deposit along the river includes a natural levee of silty material immediately adjacent to the stream, and layers of clay and silty clay in the slackwater areas adjacent to the river, at lower levels than the natural levee. The project site is located in a depressional area, which is covered with fine-grained sediments characteristic of slackwater deposits.

The marina was to be developed in a topographic low. As mentioned previously, the marina site was separated from the river by a natural levee deposit of silty soil. On the other side of this levee, the small, ephemeral stream occupied a portion of the depression. The stream cut through the natural levee materials and entered the river somewhat downstream from the major portion of the marina site. The depression which was to be changed for the marina project was bounded on the south by a series of limestone cliffs. The limestone was heavily weathered and contained many gullies and enlarged joints, with solution cavities within the mass of the rock. On level spots atop the cliffs and bluffs, the limestone was covered with a red-brown silty clay residual soil. In some areas, a thin layer of silty windblown material was found above the silty clay residual soil.

The two channels of the marina and the adjacent fill areas, for the most part, were located in a meadow environment. Ecologically, this area was classified as an upland, disturbed meadow. The area had been used intermittently for agriculture, thus disturbing its natural condition. In some spots within the project site, the meadows were replaced by groves of upland hardwood trees, consisting primarily of maples, beeches, and oaks. Ground cover in the area was fairly dense, particularly in the wooded portions. However, no known unique floral species or associations of plants were found in the area of the proposed project.

Personnel of the United States Fish and Wildlife Service had performed a field survey in the area of the proposed marina. On the basis of the survey, it was concluded that the overall wildlife values of the project area were not significant. The most numerous wildlife species in the project area were song birds. The habitat in the meadows and occasional wooded areas of the project site was capable of supporting a low to moderate population of upland game species such as quail, rabbit, and squirrel. The habitat along the ephemeral stream, particularly in the lower areas, was considered favorable for numerous amphibian and reptilian species.

Higher aquatic organisms were not found in great numbers in the project area. The overall fishery value of the project site was considered to be severely limited. Fish from the major river could conceivably enter the depressional area occupied by the creek, during periods of flooding when the creek was filled with river water. However, during most of the year, particularly during dry weather, the small size and intermittent character of the creek precluded any significant fish population in the creek. At the entrance of the creek into the major river, some spawning by river fish species could occur.

With respect to the cultural or man-made aspects of the site, no significant historical sites were found within the project area. However, archeological reconnaissance conducted on the site had uncovered scattered artifacts from archaic cultures. Similar artifacts had been discovered along the natural levee adjacent to the river, and also along the tops of the limestone cliffs to the south of the project. The occasional artifact discoveries within the project site area were not considered to represent a unique archeological resource. The cultural value of the site, with respect to land use, consisted primarily in its value as "open space" for the community. The residential community near the project site consisted of a group of families who occupied houses and camps along the riverbank adjacent to the site of the proposed marina. These families occupied the housing units in a strip along the river, on a temporary basis in some cases, and on a permanent basis in other cases. These housing units were served by a two-lane roadway located parallel to the river, along the top of the natural levee deposit.

6-7.3   Predicted Impacts of the Marina Development

After the initial plan for development of the marina complex had been completed and had been reviewed by the environmental consultants employed by the developer, a number of detrimental consequences of the development were identified. These impacts included an alteration of the topography of the site through the conversion of an open, relatively flat area of meadow and park land to a marina and housing development. The change in topography at the site had additional impacts with respect to drainage of water from the site and effects on the aesthetic quality of the site. The placement of soil excavated from the channels of the marina in the area between the channels and in the areas along the length of the channels on their outer edges changed the overall drainage pattern of the area. The project site topography, before the implementation of the marina construction plan, was such that the drainage of surface waters occurred toward the ephemeral stream in the depression. Water falling on the ground surface in the area infiltrated to a slight degree; the major portion of all precipitation, however, drained from the site as surface runoff downslope from the natural levee and from the area of the limestone bluffs into the depression and on to the river through the channel of the small creek. The construction of the marina facility would divert a portion of the drainage waters toward the residential area immediately adjacent to the river. Placement of fill along the length of the marina channel closest to the river would raise the elevation of the ground surface in that area and cause a slight amount of flow to be directed toward the river. This small amount of drainage water was to be collected in a ditch which would be constructed parallel with the marina channel, and collected water was to be diverted into the channel and, hence, into the river.

The change in topography and use of the land at this site was predicted to have relatively minor effects with respect to the wildlife on the site, because of the lack of abundant or unique species on the site prior to construction. The direct effects on fish and wildlife were considered to be rather minor, although some wildlife habitat would be destroyed. More serious effects were considered to be possible through the creation of stagnant water conditions in the marina channels. The indirect effects of the stagnant water conditions in the marina channels would result from the gradual accumulation of nutrients and pollutants in the stagnant water. Because of the presence of these nutrients and pollutants in the stagnant marina channels, health problems could be created. Additionally, objectionable odors and visual deterioration of the area could be created through such pollution of the water in the marina channels. It was considered possible that most of the fish in the channels, although they were not considered to be plentiful, would be killed if polluted water conditions developed, with consequent depletion of dissolved oxygen in the water. The major ecological impacts of the project were considered to be centered on the possibility of creating local stagnant water conditions, rather than through endangering unique species of flora or fauna or destroying unique habitat.

The danger of creating stagnant water conditions in the marina channels had been cited by the Environmental Protection Agency and by the United States Fish and Wildlife Service as a possible source of degraded water quality and associated ecological damage.

In the vicinity of the proposed marina, drinking water supplies for a residential area, located on the bluffs to the south of the project site, were obtained from a well located on the riverbank within the project area. Water from this well was used to satisfy the needs of approximately 40 families living in residences on the limestone bluffs. This situation was changed soon after the developer began to design the proposed marina complex, because the water supply utility located in the nearby metropolitan area began to furnish pipeline water to the residents in the bluff area. However, one resident of the area could not be served by this utility; as a result, he remained wholly dependent upon the riverbank well for drinking water supplies. Excavation for construction of the marina channels, as envisioned in the proposed plan of the developer, would interrupt the water supply to this single resident.

Another factor considered in the environmental assessment was the possible effects of the marina complex on the planned construction and operation of a raw water intake and treatment facility to be built a short distance downstream from the proposed marina complex, on the banks of the river. The local water utility had developed plans for the construction of an intake and an associated treatment facility some time before the developer of the marina complex began to design his facilities. The pollution control regulations of the state in which the marina complex was to be built included a provision stating that no primary or secondary wastewater treatment plant effluent could be discharged into any public water supply at a distance of less than 1,000 feet downstream from the water supply intake structure, or at a distance more than 5 miles upstream above such an intake. The discharge from a wastewater treatment plant could be placed closer to the public water supply intake only if the wastewater treatment plant effluent had been subjected to tertiary treatment. The original plan for the marina complex called for secondary treatment of wastewaters in the treatment plant designed to serve the housing units on the site and the boats to be docked at the marina facilities. The discharge of wastewaters from this secondary treatment plant could not be permitted at the location originally chosen by the developer, because of the regulation mentioned above.

The social impacts of the proposed facility were considered to be related primarily to the construction of housing units on the site. The ultimate housing capacity of the 400 townhouse units and the 200 apartment units to be built on the site was considered to be approximately 1800 persons. The proposed development corresponded to a housing density of eight dwelling units per acre. The local zoning regulations in the area of the project permitted a density of 30 living units per acre, and the zoning regulations in adjacent areas permitted as much as 12 living units per acre. However, the increase in the population of the project site, as a result of construction and occupation of the townhouses and apartments in the complex, was considered to have major effects on the service capabilities of the regional government in the area. For example, the narrow, poorly maintained secondary road was designed to serve the needs of the 50 families living along the river slope of the natural levee adjacent to the proposed marina site. The influx of 1800 persons as occupants of the marina complex would generate traffic which would seriously overload this roadway. Also, the permanent occupancy of some of the townhouse units would lead to a need for increased school facilities for the children living in the marina complex housing units. Finally, an increase in the permanent population of the project area as a result of construction of the marina complex would lead to greater demands for police protection, fire protection, and medical services in this portion of the region, as compared to the needs of the existing population near the project site. These detrimental social consequences were identified by residents of the riverside community and were propounded vigorously in private correspondence with the federal agency which was to grant the permit for the project, as well as in local media and in public hearings.

The construction of the marina complex was considered to have possible beneficial effects on these same existing residents of the riverside development since these residents depended upon well water for drinking water supplies prior to the construction of the marina complex. Since water lines were to be constructed to the marina complex by the regional water supply utility, the residents of the riverside community could obtain similar service from such a pipeline. Additionally, these residents could obtain access to gas pipelines constructed to serve the marina complex. These two impacts on the existing community at the project site were considered to be significantly beneficial.

In addition to the impacts previously listed, a significant possible impact was identified with respect to the groundwater flow conditions at the site of the proposed marina. Most of the residents in the riverside community along the river slope of the natural levee depended upon wells excavated into the sand and gravel beneath the surface for drinking water and utility water supplies. The excavation of the marina channels conceivably could rearrange the groundwater flow patterns in the project area to affect the available groundwater supply in these wells. The principal source of groundwater in the aquifer beneath the project site is the river itself. Excellent hydraulic connections exist between the river and the underying sand and gravel layers. Additional recharge to the underlying aquifer occurs in this area through seepage of precipitation through the upper soil layers and into the aquifer. Very little water flows into the aquifer from the limestone bluff area to the south of the project site, because of the bedrock topography and because of the cavernous nature of the limestone.

The creek which flowed through the depression which was to be transformed into the marina channels served as a temporary recharge and outlet area for groundwater. During periods of high river elevation, groundwater reserves would be recharged through the banks of the small stream at the site. During periods of low flow, groundwater would seep into the channel of the ephemeral stream to supplement surficial flows. Excavation of the marina channels could conceivably have shifted the groundwater levels slightly so that the water level in wells along the natural levee area could have been lowered slightly. However, the lowering of water levels in these wells would not have been significant in any case.

During the course of the environmental assessment, a number of other adverse impacts were identified, which are too numerous to describe in detail here. Some of these impacts can be listed, however: loss of open space and semi-wilderness area; deterioration and interruption of attractive scenery; increase in noise levels in the area through increase in boat traffic in the marina; and a disruption of the cultural patterns of the residents of the riverside community along the natural levee at the site of the proposed facility. This last item does merit some extended description since this was the most important single social impact identified by the members of the assessment team. The riverside community consisted of approximately 50 households wherein the residents occupied their facilities on either a temporary basis or a permanent basis. Opposition to the proposed development was completely unanimous among the residents of this community. One of the principal reasons for their purchase of land and housing at this location was a desire to escape the pressures of urban life in the nearby metropolitan area. Consequently, any development which posed the possibility of increasing population density, increasing noise, or displaying any of the signs of urbanization was firmly opposed by these individuals. These people drafted letters and petitions protesting the development and submitted these petitions to local lawmakers as well as to the federal agency which would grant or deny the permit for the proposed activity. Most of the letters and verbal comments given by the residents of the area, in opposition to the proposed development, described several categories of undesirable effects of the proposed action: undesirable increase in population density in the area; overloading of inadequately maintained access roads; overcrowding in school facilities; loss of rural, isolated atmosphere; increase in the noise level in the general area; possible contamination of water supplies; overloading existing inadequate police and fire protection services; destruction of wildlife area; improper drainage of surface waters from the elevated fill areas of the proposed development; and the possibility that the project would not be completed and would become a permanent eyesore and detriment to the area.

In contrast to the feelings of the residents in the riverside community adjacent to the proposed project area, a large number of boating enthusiasts who utilized the river in this locality for recreational purposes expressed enthusiasm for the proposed development. A large number of letters were submitted to the federal agency in connection with this project, supporting the proposed development of a marina and housing complex. The principal benefits cited for such a development were an increase in the number of available boat docking facilities, which were relatively scarce in the metropolitan area bordering the river, and the development of a year-round recreational facility through the provision of permanent housing in connection with boat docking facilities.

A number of the impacts identified by the consultants were considered to be unavoidable adverse effects of the proposed project.

For example, development of the marina complex would produce a permanent loss of habitat and ground cover for the animals which were living in the meadows and wooded areas of the site. The scenic view of the undisturbed site from the river bluff would be disrupted through the construction of the housing units and the marina channels. Obviously, a significant loss of open space in the area of the project would be caused. The essentially rural atmosphere of the project site, as it existed prior to development of the marina, would be drastically altered with an increase in traffic, noise level, and many of the other external effects of urbanization. Relatively minor impacts would be felt on archeological resources in the area of the project site. The principal detrimental effect which could not be avoided if the project were to be built consisted of the adverse social and psychological effect of the project upon the present residents of the riverside community along the natural levee adjacent to the site.

6-7.4 Alternatives to the Proposed Action

After the adverse environmental impacts associated with the proposed plan of development had been identified by the consultants, they suggested to the developer several measures which could be taken to mitigate the undesirable impacts of the proposed activity. These alternative actions were designed to mitigate impacts in three particular problem areas. First, the development of stagnant water conditions in the marina channels could cause a significant accumulation of pollutants and nutrients through runoff from adjacent lands and through pollutant introduction from the effluent of the proposed wastewater treatment facility. This accumulation of pollutants could lead to depletion of dissolved oxygen in the water of the marina channels, with resultant killing of all fish in those waters. Additionally, significant problems of odor and visual insult could result through such stagnant water conditions. To alleviate this possible problem, the consultants suggested the introduction of river water into the marina channels. The developer agreed to withdraw water from the river, pass it through a gravity aerator, and pass it through both of the marine channels at a rate of approximately 2 million gallons per day. This pumping rate corresponded to a total turnover time, for the water in the marina channels, of approximately 17 days. Pumping 2 million gallons per day of aerated water through the marina channels was considered sufficient to prevent stagnation, with its consequent buildup of natural and man-made pollutants and the deterioration of water quality in the channels.

The consultants also developed an alternative plan to prevent stagnation through the installation of an aeration system constructed on the bottoms of the two marina channels. The oxygen flow in this system could be regulated to provide a minimum of 4 ppm of dissolved oxygen in the marina channel waters at all times. Additionally, a small connecting channel could be created to increase the circulation of the water between the two channels. Maintenance of a dissolved oxygen level of at least 4 ppm in the marina waters at all times was considered sufficient to prevent any significant euthrophication.

The consultants also recommended that the wastewater treatment facility plant for the site be upgraded to provide tertiary treatment, rather than secondary treatment. In this way, the wastewater treatment plant effluent could be discharged to the river, as planned, without causing any hazard to drinking water supplies, obtained in the proposed raw water inlet to be constructed at a short distance downstream from the marina complex. The consultants suggested that the effluent from the tertiary treatment plant could be discharged into the marina channels to promote circulation of the channel waters. The effluent from the tertiary treatment plant, in all likelihood, would be of higher quality than the water in the river near this site, and thus, discharge of this effluent into the marina channels would yield a resultant water quality higher than the quality of the water in the river adjacent to this site.

The problem of possible contamination of the water in the wells utilized by the residents of the riverside community at the site caused serious concern. The consultants predicted that the construction of the marina channels could possibly lower the groundwater level slightly, in the wells used by the present residents of the riverside community. No impairment of the water quality in those wells was anticipated, however. If the developer elected to install an aeration system, as suggested by the consultants, to prevent stagnant water conditions, the water in the marina channels was considered to be of the same quality as the water in the river channel itself. Any migration of this water into the wells used by residents in the area would not pose any pollution problem or hazard. State regulations to control water pollution required retention tanks and other pollution control devices on all boats having head and galley facilities. For this reason, if regulations were implemented and strict compliance was maintained, no discharge of pollutants from the pleasure craft in the marina should occur. The wastewaters from the pleasure craft in the marina could be taken to the wastewater treatment facility at the site and rendered suitable for discharge into the river or into any other large body of receiving water.

To further guarantee that no contamination of the aquifer feeding the water wells at the site would occur, the consultants suggested the installation of a clay liner on the surfaces of the marina channels. The developer subsequently agreed to install such a clay liner on the channel bottoms, to form an impervious layer which would prevent exfiltration of the water from the aquifer underlying the area, and also prevent infiltration of marina water into the aquifer. During periods of low river level, a potential would exist for seepage of water from the proposed marina channels toward the river and the well field utilized by residents of the riverside community. A consequent possibility of well contamination would exist in this instance. This migration of polluted waters into wells could be effectively prevented by the installation of a sufficient thickness of clayey soil along the channel bottoms. Seepage could be retarded sufficiently so that during the entire low river stage season, the seepage rate would be so low that no significant flow of water would occur from the marina channels into the aquifer beneath the site. Locally available clay soils in the vicinity of the site were known to have coefficients of permeability on the order of 10^{-4} feet per day to 10^{-6} feet per day, after compaction. If such low permeabilities were obtained in the clay liners constructed on the bottoms of the marina channels, the possibly polluted waters in the channels would migrate only several inches through such a clay liner during the entire dry season of the year. A minimum thickness of at least 3 feet of clay liner was recommended as protection against accidental dredging or cutting in the channels, and the complex developer agreed to this recommended measure.

The possible contamination of wells in the area was considered to be a rather unlikely occurrence since the aquifer material into which the wells penetrated in this area was a well-graded sand. This sand constitutes an excellent filter medium. A minimum distance of approximately 300 feet was found between the wells adjacent to the site and the marina channels. Consequently, at least 300 feet of well-graded sand would be available to act as a filter for any water leaking from the marina channels into the aquifer and hence into a well. The permeability of the sand aquifer was such that water from the marina could migrate through the sand at the rate of 3 to 4 feet per day, if it had penetrated the clay lining to be installed in the channels. For all practical purposes, the sandy characteristics of the aquifer were considered sufficient to provide excellent filtering capabilities for any water entering from the marina channels into the aquifer.

The disruption of drinking water supplies for one of the residents of the area, and for the 40 homes located on the limestone bluffs to the south of the project area, has been mentioned above. The problem of supplying drinking water to the homes on the bluffs was alleviated through the revision of the developer's plans to re-route the existing water supply pipes, so that the drinking water supplies for all residents would not be interrupted during any phase of construction or subsequent operation of the marina facilities.

6-7.5 Other Aspects of the Project

The logic of the environmental impact assessment was utilized to consider the proposed marina development. This logic included a determination of the relationship between local short-term uses of the environment and the maintenance and enhancement of long-term productivity in the area. In this connection, certain trade-offs were identified. The construction of the marina complex would create a permanent loss in available open space, which is a valuable commodity in an area near a crowded metropolitan region. The agricultural potential of the site also would be lost, although agricultural utilization of the site had not occurred for a number of years. The potential utilization of wildlife resources in the project area would be reduced inevitably by the proposed construction of the marina, but the effect of this disruption must be evaluated against a background of increasing urbanization of the total area and consequent disruption of wildlife habitat in the long run.

The proposed marina complex was considered to offer a beneficial long-term potential, in that it would afford residents of the metropolitan region the opportunity to experience a rather unique type of residential living. The residents of the townhouses in the marina complex would live above the 100-year flood stage, would be serviced by all of the utilities available in the nearby metropolitan region, and would have at their disposal year-round boat docking facilities, immediately adjacent to their place of residence. This provision of recreational facilities, in the form of docking facilities, was considered to be significant since such facilities were relatively scarce along the river adjacent to the metropolitan region.

Irreversible and irretrievable commitments of resources associated with the project were also considered. The most significant irreversible commitment of resources included in the project would be the loss of approximately 100 acres of wildlife habitat. This area served as a source of food and cover for wildlife prior to construction of the project, and this resource would be lost with construction of the proposed marina. No other significant irretrievable commitments of resources were identified for the project.

6-7.6 Subsequent Developments

After the consultants recommended certain revisions to the original plan, the developer of the marina complex agreed to incorporate such revisions into the final plan for the project. In order to present the original plan and the suggested alternatives to the public, the developer requested the federal agency involved to schedule a public hearing and to invite all interested citizens in the area. Such a hearing was held at a school facility near the project site. Approximately 250 persons appeared at this hearing, and the majority of the persons at the hearing expressed serious reservations concerning the project, and many expressed outright hostility to the proposed development. A number of questions were raised concerning increases in noise levels associated with the proposed project, deterioration of water quality as a result of the proposed construction, overloading of service facilities, such as fire protection and police protection, and a number of other controversial issues concerning the possible social effects of the proposed development. With respect to the technical characteristics of the project, a significant objection was raised during the public meeting, by a resident of the riverside community. This individual cited the dependence of the developer on the wastewater treatment plant to alleviate many of the problems identified originally by the consultants. This citizen pointed out that any processing plant or similar facility inevitably must experience some downtime, and the quality of its operation is dependent upon the human operators who supervise and control operations at the plant. Failure to operate the treatment plant correctly could produce serious pollution problems in the marina. This point was addressed in the revised environmental assessment, which later became the basis for an environmental impact statement on the proposed facility.

6-7.7 Summary

This case serves to illustrate the possibilities for remedial action during design which can be obtained through early identification of detrimental impacts of a proposed activity. It also illustrates that the overall methodology of impact assessment can be applied to relatively small projects. The inclusion of recommended improvements in design in the final plans for this facility represented a very positive beneficial effect of the assessment process. Additionally, the assessment provided a good basic tool for discussion and for eliciting opinion from residents of the area during the public meeting. The presentation of the assessment by consultants to the developer at the occasion of the public meeting served as a convenient focus for later comments and criticisms. Many of the comments and criticisms which were raised by the residents of the area were answered adequately by the investigators who prepared the environmental assessment. However, several pertinent statements were made by members of the public during the public hearing and these statements also brought about changes in the basic design wherever possible. This close cooperation of the public in the design and planning process is one of the most desirable goals to be sought in any environmental assessment effort.

6-8. SUMMARY

In this chapter, a number of case histories have been presented to illustrate the application of environmental assessment technology to a variety of proposed activities. The projects described on the preceding pages ranged from developments involving rather small expenditures of resources and money, to major projects involving very significant commitments of resources. The environmental assessments for each of these cases, however, were based upon the same general system of investigation and evaluation.

One of the most important benefits to be obtained in any environmental assessment project is the identification of detrimental effects anticipated from a proposed activity, and the opportunity to modify the proposed plan of action to mitigate the identified detrimental effects. If all viable alternatives are examined in a comprehensive fashion, it is possible not only to identify areas where changes in particular projects can eliminate or reduce undesirable impacts, but it is possible to evaluate the relative merit of cancelling all plans for the proposed activity. The assessment procedure, as outlined in these cases, furnishes information to the public for use in making decisions concerning the allocation of resources to preserve and improve environmental quality. A comprehensive assessment will allow laymen and professionals to make informed estimates of the relative costs of proceeding with proposed activities or with alternative schemes.

One of the most serious misunderstandings concerning environmental assessment activities is the mistaken notion that completion of an environmental impact assessment will virtually guarantee the abandonment or indefinite postponment of a contemplated activity. Much of the publicity which has been devoted to the subject of environmental impact assessments and statements has been directed toward emphasizing the dramatic effects of construction stoppages resulting from litigation based upon environmental impact assessments or statements. This negative publicity has produced an unfortunate impression in the minds of many persons concerning the true purpose of environmental impact studies. If the impact assessment is conducted in a proper fashion, it will serve as a very important tool for analysis of possible impacts, and as a means of identifying possible changes in design or construction techniques which could eliminate or reduce undesirable effects from a particular activity.

Finally, properly conducted environmental impact assessment studies, as illustrated in the case histories presented in this chapter, are very important tools for involving members of the public in governmental decision making. The stated purpose of the National Environmental Policy Act of 1969 was to provide a vehicle for establishing a national policy on environmental conservation and protection. It is impossible to establish a national policy without the participation of the general public. The systematic assessment and evaluation of proposed activities and their possible effects on the environment provides a means for achieving meaningful public participation in formulating overall environmental policies.

7

Summary

A number of very important concepts and ideas have been presented in this text. These ideas are considered, by the authors, to be very important in achieving effective environmental protection and conservation. They are also vital to the most efficient and wise use of resources in improving the quality of life. Because of their importance, it is appropriate to briefly review and summarize the material and concepts presented in the preceding chapters.

7-1. WHY ASSESS THE ENVIRONMENT?

Many of the readers of this book will have a variety of reasons why they want to engage in, or examine, assessments of environmental quality and environmental impact. In the first chapter of this text, examples of man's unwise use of natural resources were given to furnish some additional reasons for developing assessment procedures and for engaging in such activities. These techniques have been applied to classes of problems in the United States and in other countries, primarily in evaluating the effects of a particular physical project development on the biological and physical environment. This scope of assessment is too narrow. Almost all facets of human activities could be subjected to systematic evaluation and assessment as described in this text. Obviously, this would hardly be practicable. However, the techniques and methods described in this text can be applied to the evaluation of situations which have great political, economic, and social significance. All too often, in past efforts, environmental impact assessments and studies have consisted of voluminous inventories of biological and physical data, with accompanying speculation on the effects of proposed activities on flora and fauna in the region of the proposed facility or activity. The effects of particular projects on communities and social systems have not received proper attention. Likewise, major policy decisions taken by members of the federal government have not been subjected to the same sort of scrutiny as have federal agency actions which could have an effect on the physical environment. Because of the obvious connection between interstate commerce regulations, domestic tax structures and other policy issues, and consequent environmental effects, the lack of comprehensive evaluation of these primary causal activities must be remedied.

Of course, a great number of examples could be cited to show that, even if the term "environmental quality" is restricted primarily to physical and biological conditions in the environment, comprehensive evaluation and assessment of probable effects of man's activities, in most cases, has not been undertaken prior to the implementation of the proposed plans and designs. A number of examples were presented in the first chapter of this book, to show the ill-effects of engaging in major actions which have significant impacts on the environment without previous evaluation of those effects. These examples were presented in hopes that they would illustrate sufficiently the need for comprehensive and thorough environmental assessments and evaluations.

7-2. ENVIRONMENTAL PROTECTION, BY LAW

Early efforts to improve the quality of life in this country were directed at eliminating sources of pollution (and similar detrimental consequences of industrialization) through reliance on the provisions of common law. When these provisions did not prove to be adequate to the task of ensuring environmental quality, the nation's lawmakers devised appropriate statutes to preserve and protect various segments of the environment. Improvement in air quality was addressed at an early stage. Later, comprehensive legislation was passed in an effort to protect and improve the quality of the nation's streams and lakes. Finally, in recent years, other facets of environmental protection have been considered, with the resultant passage of federal laws dealing with solid waste management, resource recovery, and noise abatement.

In the second chapter of this text, the effort to achieve environmental quality through provisions of the common law and through statutory law was described in considerable detail. This comprehensive description of the legal framework for environmental conservation was given in order to provide a complete background to the current situation in the United States with respect to environmental preservation and protection, and also to provide a more comprehensive base for the application of the environmental assessment methodology and techniques which are described in later chapters. The intention, in this effort, was to develop a foundation of legal concepts which would define environmental standards and which would describe the individual's right to environmental quality. In many cases, adherence to a particular law or statute cannot be ascertained unless a comprehensive environmental monitoring and evaluation program is undertaken for the particular case under study. For example, the determination of the existence of a nuisance, as defined under common law, may require comprehensive sampling and testing of environmental quality in a particular area, or testing for a particular parameter which is a measure of an environmental attribute. In other words, decisions on the legality of actions which affect environmental quality can be based upon the legal framework described in Chapter 2, and upon the evaluation and assessment framework described in later chapters. This connection of laws protecting environmental quality, with the technology to evaluate environmental quality and impacts, is one of the most important themes presented in this book.

7-3. THE NATIONAL ENVIRONMENTAL POLICY ACT

The National Environmental Policy Act has emerged as the most decisive single piece of legislation in the field of environmental protection and conservation in the United States. Consequently, it is important to understand the background of this law. Considerable space in Chapter 3 was devoted to describing the origin and growth of the National Environmental Policy Act through rather tortuous legislative channels. Because this piece of legislation has had such far-reaching effects and such wide application, it is important to understand how and why it was developed.

One of the recurrent themes which appears in early chapters of this book is the interpretation and enforcement of the provisions of the National Environmental Policy Act by federal judges. The reader will recall many instances where particular landmark court cases were cited. The role of the federal judiciary in implementing and enforcing the National Environmental Policy Act cannot be over-emphasized. The judicial branch of the government has acted to fill a void created by the negligence of the executive branch. No executive office or agency appeared to be willing to act as the implementor or enforcer of the provisions of NEPA. In the early days of the legislative history of NEPA, considerable speculation was devoted to the idea that the Office of Management and Budget would be the prime enforcing agency for the forthcoming law. However, this agency failed to take such a leadership role and no other federal agency or office accepted the challenge. As a consequence, the federal judiciary entered the picture, in response to a number of lawsuits brought by environmentalists and conservation groups, in order to clarify the intent and purpose of the law.

In general, judicial interpretation of the provisions of NEPA has been literal and extremely comprehensive. Federal agencies have been required to adhere to the letter of the law, not merely to the spirit of the statute. Because of the importance of the judicial role in enforcing and interpreting NEPA, considerable attention has been devoted in this book to describing the results of landmark decisions made by federal judges. As a consequence of these important decisions, environmental assessments and impact statements have grown to be more comprehensive and thorough, even to the point of being unwieldy and cumbersome. Recently, however, new court decisions have called for the preparation of impact assessments and statements which are concise and easily understood by the public. These recent rulings appear to have brought NEPA full cycle, back to the original intention of Senator Jackson and the other lawmakers who developed the original act.

In the third chapter, the provisions of NEPA were described in detail. This detail is justified since this law was the first major, coherent statement of environmental policy for the United States. Also, this law established the Council on Environmental Quality to serve as an advisory body to the President, to guide federal policy in an effort to preserve environmental quality and protect natural resources. However, the most highly publicized and well-known section or provision of the National Environmental Policy Act was Section 102(2)C, the so-called Environmental Impact Statement Section.

7-4. THE ENVIRONMENTAL IMPACT STATEMENT

The growth of the environmental impact statement, from the requirements contained in Section 102(2)C of NEPA, was described in Chapter 4. Environmental impact statements are required for all actions involving federal agency funding or jurisdiction which are judged to have significant impact on the environment. Obviously, the terminology "significant impact" contained in NEPA is open to interpretation. One of the most serious controversies which surrounded the preparation and review of environmental impact statements in the months immediately following the passage of the National Environmental Policy Act, was the determination of the significance of predicted impacts of proposed actions. This evaluation and interpretation of identified impacts remains the most crucial issue in environmental assessment efforts to date. However, the majority of the federal agencies which are charged to prepare environmental assessments, now prepare such assessments on virtually any activity in which they are engaged. In other words, the lower threshold for significance of impact from an agency action has been reduced, by most federal agencies, to the point that virtually every agency action will be required to be described and evaluated in an environmental impact statement.

In Chapter 4, the requirements for preparation of an impact statement were described; the use of an interdisciplinary team in preparing the impact assessment and statement was emphasized; the time frame for preparation of a statement and for review of draft statements was described, as well.

The items which must be covered in any Environmental Impact Statement, prepared under the requirements of the National Environmental Policy Act, also were described in Chapter 4. In every EIS, the quality of the environment in a particular area where an activity is proposed to take place must be described in comprehensive fashion and in significant detail. This description of environmental quality must be quantitative to as great a degree as possible. Obviously, some aspects of environmental quality cannot be described in quantitative fashion, since any description of certain environmental attributes must be given in subjective terms. This is particularly true for aesthetic and social factors in the environment. However, every effort must be made to thoroughly characterize the environmental setting both with and without the proposed project or facility.

In all Environmental Impact Statements, the proposed activity under study must be described in great detail. This requirement is one of the most important features of NEPA and its requirements for impact assessment and evaluation. Much of the difficulty in achieving comprehensive environmental planning with public participation has arisen in the past through ignorance of the character of proposed activities, not only on the part of the public, but also on the part of would-be impact evaluators. Developing a comprehensive description of the proposed activity is extremely useful in that it brings project initiators into intimate contact with impact evaluators, if the project description is prepared in the proper fashion as a cooperative effort between these two groups.

It is obvious that all identified impacts of a proposed action must be evaluated and described in detail in any environmental impact assessment or statement. However, particular attention must be given to those impacts which cannot be avoided if the proposed action is implemented. One of the features of comprehensive environmental evaluation which has been neglected in this regard is the identification of the benefits to the environment which may accrue from implementation of a proposed action or facility. In order for any assessment to be comprehensive and complete, all impacts, both beneficial and detrimental, should be fully described and evaluated.

The requirements of NEPA include a stipulation that other aspects of environmental impact must be described. The relationships between short-term uses of the environment and long-term maintenance of productivity and quality of the environment must be described. Also, any irretrievable commitments of resources, both human and natural, must be identified and fully evaluated. Finally, the relationship of the proposed activity with the land use plans developed for the area in which the activity is to take place must be described and evaluated.

A properly prepared environmental impact statement can be a very valuable tool in achieving effective environmental planning. However, many of the impact statements which have been prepared in recent years have proven to be ill-conceived, cumbersome efforts. Achieving effective environmental planning through the preparation of comprehensive and accurate environmental impact statements cannot be accomplished unless a systematic approach is used in the preparation of these assessments and statements. A methodology for this work must be developed.

7-5. ASSESSMENT METHODOLOGY

In Chapter 5, the methods of impact assessment and the techniques for organization of assessment personnel were described. Many of the traditional methods for the evaluation and characterization of engineering plans and designs have proven to be ineffective in environmental assessment activities because of the necessity, in these traditional methods, to characterize various attributes of the environment in a quantitative fashion. In technical and economic evaluations, all too often a point is reached at which a dollar value or a number value must be assigned to a particular facet of environmental quality which cannot be described quantitatively. For this reason, traditional engineering and economic analyses have not been effective in assessing environmental impact.

Graphical techniques have been employed with a certain measure of success in identifying particularly fragile segments of the environment and in describing the physical system in an area under study. Also, graphical techniques have been used successfully to portray the severity of impact in a particular area, for a particular segment of the environment, in a qualitative fashion. In future environmental impact assessment efforts, graphical techniques are likely to be most effective as means of portraying the location and severity of particular impacts. However, they cannot be relied upon to furnish a quantitative estimate of impact severity or character.

A number of other techniques have been developed by various individuals and organizations, in efforts to devise a systematic and universal approach to impact assessment. Matrices have been developed by a number of organizations. In these matrices, characteristic activities which may be included in a proposed action are listed on one axis, and possible impacts, or areas of environmental quality which may be affected, are listed on the other axis. Interactions between particular activities and possible areas of impact thus may be displayed through the use of such a matrix. Matrices have been developed by some individuals with the intention of providing a quantitative assessment tool. In general, these efforts have been unsuccessful. Matrices remain a useful way to portray areas of impact, but they are not effective as quantitative analysis tools.

Checklist methods of environmental assessment have also been developed by a number of investigators. In these techniques, a particular project is compared, for possible areas of impact, with long lists of environmental attributes or possible areas of environmental impact. These methods serve primarily to ensure that no possible impact is neglected, or that no segment of the environment is not evaluated. However, the lists which are used as memory aids in this effort may not include all conditions encountered at a particular site for a particular proposed activity. Thus, checklist methods also are not foolproof.

A number of quantitative evaluation schemes have been developed for environmental assessment. An outstanding example of such a quantitative system is that developed by the personnel of the Battelle Memorial Laboratories, for use in evaluating the impact of water resources projects and activities. In general, most quantitative methods are not sufficiently flexible to be applied to all kinds of projects in all geographic locations. The quantitative nature of these methods indicates that value judgments have been made concerning the importance of particular parameters and the importance of certain degrees of impact, by the method developers, in the perfection of the quantitative technique. Thus, any quantitative method can be questioned on the basis that these value judgments may not be appropriate for the particular case under investigation.

A general approach to environmental assessment was suggested in Chapter 5. Matrices were recommended as useful tools in preliminary identification of particularly fragile segments of the environment and in preliminary identification of major impacts. The development of a checklist by a committee representing project initiators, impact evaluators, and the general public, for use in examining the effects of a particular project in a specific locality, also was recommended. In this regard, the authors feel that much duplication of effort in impact assessment could be eliminated if classes of projects could be identified and general environmental assessments could be prepared for these classes of projects or activities. In other words, if particular types of activities, such as the construction of nuclear power plants, could be examined by a panel of experts representing, perhaps, the nuclear power industry, environmentalist groups, consulting engineers, federal agency personnel, and the general public, a general assessment of impact from such facilities could be developed. For any subsequent proposed nuclear power development, the general impacts of such a project could be easily identified. Then, particular attention could be directed toward the specific location in which the plant was to be built and toward the specific characteristics of the plant itself. The time, effort, and money expended in the environmental assessment could be spent more profitably in examining the particular situation, rather than in discussing, once again, the general situation in the entire country.

In the second part of Chapter 5, methods of personnel organization were described. In formulating an environmental assessment team, the most appropriate method of identifying team members is to identify particular tasks in the assessment effort and then to determine who are the most appropriate individuals to accomplish those tasks. However, the assessment cannot be prepared by a number of individuals working independently. NEPA requires the use of an interdisciplinary team in the preparation of impact assessments and statements. In impact studies not required by NEPA, but required by other laws, either state or federal, common sense would require the use of such an interdisciplinary team. In formulating this team and in achieving effective cooperation among team members, one of the most important functions is that of the team leader or project manager. This individual must be chosen carefully for his abilities to develop close cooperation among individual team members, and for his breadth and scope of knowledge in all the various disciplines represented by the members of the assessment team. This individual must establish contact with members of the general public at an early stage in the assessment process. He must also coordinate the efforts of specialists in public relations, graphic arts, and other related activities which are necessary for the preparation of clear and concise impact statements.

7-6. CASE HISTORIES

In Chapter 6, a number of case histories were described to illustrate the use of systematic environmental assessment technology, and to illustrate the improper approach to environmental assessment. Two cases involving highway construction were described. In one case, inadequate involvement of the public and insufficient evaluation of alternatives doomed the environmental assessment to failure, and effectively stopped the project. In the other highway case history described in this chapter, the widening of an existing highway caused rather unique impacts on individuals who were somewhat accustomed to the presence of such a facility. The development of a plan for construction and operation of waste-water management facilities in a major metropolitan area was a third case history described in Chapter 6. The use of comprehensive planning techniques in the development of this plan served to facilitate the preparation of an environmental impact assessment and statement on the plan. The use of a citizen advisory committee to quantitatively and comprehensively evaluate alternatives was a very significant feature in the preparation of this plan. In effect, much of the work involved in the impact assessment was accomplished during the development of the plan for which the impact assessment was prepared.

In another case history in Chapter 6, the use of systematic environmental assessment methodology in the evaluation of small-scale projects was illustrated through the examination of disposal methods for construction site-clearing wastes. Several alternative methods, including uncontrolled open burning, controlled open burning, use of satellite incinerators, use of a central incinerator, and use of landfills, were examined through the application of assessment methodologies. This case illustrates the potential for use of comprehensive assessment methodologies in evaluating land use plans and activities on a small scale.

A major federal activity having a significant effect on a segment of the environment was described in another case history in Chapter 6. The construction of a navigation lock and dam on a major river in the midwestern United States was evaluated for environmental impact. This illustrated an unusual circumstance; in this particular instance, the action which was evaluated was virtually complete at the time the evaluation was undertaken. In many respects, the prediction of impact, in this case, was facilitated, since many of the impacts were obvious to the project evaluators. However, since a series of navigation locks and dams were to be built along this same river, the results of the impact analysis for this virtually complete facility were very valuable in accomplishing the impact assessments for the remaining structures.

The final case presented in Chapter 6 described the use of impact assessment methodology in evaluating the proposed construction of a marina complex. The major issue, in this case, was the effect on a small recreational and vacation community, which would be disrupted through construction of the marina complex. The use of assessment methodology allowed the consultants for the marina developer to identify potential impacts and to suggest means to mitigate those impacts. Detrimental consequences of the proposed activity were not totally eliminated, but the overall impact of the project was considerably reduced. Additionally, the use of impact assessment methodology served to develop a much more concise and lucid description of the proposed development than had been available prior to the assessment study.

7-7. CONCLUSION

The authors have undertaken this book in hopes that it would provide valuable assistance to persons who are interested in determining the quality of the environment in which they live, and in evaluating the effects of proposed activities. No system or approach to any activity is foolproof. Obviously, the methods and techniques which are suggested in this volume are not perfect or foolproof. However, it is hoped that the application of the techniques described herein, with a knowledge of the importance of impact assessment and an awareness of the legal basis for environmental protection, will permit the citizens of this country and the world to make well-informed decisions on the use and allocation of resources, and thus, to improve the quality of life for present and future generations.

Appendix A

National Environmental Policy Act of 1969

NATIONAL ENVIRONMENTAL POLICY ACT
OF 1969

Be it enacted by the Senate and House of Representatives of the United States of America in Congress assembled, That this Act may be cited as the "National Environmental Policy Act of 1969."

PURPOSE

SEC. 2. The purposes of this Act are: To declare a national policy which will encourage productive and enjoyable harmony between man and his environment; to promote efforts which will prevent or eliminate damage to the environment and biosphere and stimulate the health and welfare of man; to enrich the understanding of the ecological systems and national resources important to the Nation; and to establish a Council on Environmental Quality.

42 U.S.C. 4321

TITLE I

DECLARATION OF NATIONAL ENVIRONMENTAL POLICY

42 U.S.C. 4331   SEC. 101. (a) The Congress, recognizing the profound impact of man's activity on the interrelations of all components of the natural environment, particularly the profound influences of population growth, high-density urbanization, industrial expansion, resource exploitation, and new and expanding technological advances and recognizing further the critical importance of restoring and maintaining environmental quality to the overall welfare and development of man, declares that it is the continuing policy of the Federal Government, in cooperation with State and local governments, and other concerned public and private organizations, to use all practicable means and measures, including financial and technical assistance, in a manner calculated to foster and promote the general welfare, to create and maintain conditions under which man and nature can exist in productive harmony, and fulfill the social, economic, and other requirements of present and future generations of Americans.

(b) In order to carry out the policy set forth in this Act, it is the continuing responsibility of the Federal Government to use all practicable means, consistent with other essential considerations of national policy, to improve and coordinate Federal plans, functions, programs, and resources to the end that the Nation may—

(1) fulfill the responsibilities of each generation as trustee of the environment for succeeding generations;

(2) assure for all Americans safe, healthful, productive, and esthetically and culturally pleasing surroundings;

(3) attain the widest range of beneficial uses of the environment without degradation, risk to health or safety, or other undesirable and unintended consequences;

(4) preserve important historic, cultural, and natural aspects of our national heritage, and maintain, wherever possible, an environment which supports diversity and variety of individual choice;

(5) achieve a balance between population and resource use which will permit high standards of living and a wide sharing of life's amenities; and

(6) enhance the quality of renewable resources and approach the maximum attainable recycling of depletable resources.

(c) The Congress recognizes that each person should enjoy a healthful environment and that each person has a responsibility to contribute to the preservation and enhancement of the environment.

SEC. 102. The Congress authorizes and directs that, to the fullest extent possible: (1) the policies, regulations, and public laws of the United States shall be interpreted and administered in accordance with the policies set forth in this Act, and (2) all agencies of the Federal Government shall— 42 U.S.C. 4332

(A) utilize a systematic, interdisciplinary approach which will insure the integrated use of the natural and social sciences and the environmental design arts in planning and in decisionmaking which may have an impact on man's environment;

(B) identify and develop methods and procedures, in consultation with the Council on Environmental Quality established by title II of this Act, which will insure that presently unquantified environmental amenities and values may be given appropriate consideration in decisionmaking along with economic and technical considerations;

(C) include in every recommendation or report on proposals for legislation and other major Federal actions significantly affecting the quality of the human environment, a detailed statement by the responsible official on—

(i) the environmental impact of the proposed action,

(ii) any adverse environmental effects which cannot be avoided should the proposal be implemented,

(iii) alternatives to the proposed action,

(iv) the relationship between local short-term uses of man's environment and the maintenance and enhancement of long-term productivity, and

(v) any irreversible and irretrievable commitments of resources which would be involved in the proposed action should it be implemented.

Prior to making any detailed statement, the responsible Federal official shall consult with and obtain the comments of any Federal agency which has jurisdiction by law or special expertise with respect to any environmental impact involved. Copies of such statement and the comments and views of the appropriate Federal, State, and local agencies, which are authorized to develop and enforce environmental standards, shall be made available to the President, the Council on Environmental Quality and to the public as provided by section 552 of title 5, United States Code, and shall accompany the proposal through the existing agency review processes;

(D) study, develop, and describe appropriate alternatives to recommended courses of action in any proposal which involves unresolved conflicts concerning alternative uses of available resources;

(E) recognize the worldwide and long-range character of environmental problems and, where consistent with the foreign policy of the United States, lend appropriate support to initiatives, resolutions, and programs designed to maximize international cooperation in anticipating and preventing a decline in the quality of mankind's world environment;

(F) make available to States, counties, municipalities, institutions, and individuals, advice and information useful in restoring, maintaining, and enhancing the quality of the environment;

(G) initiate and utilize ecological information in the planning and development of resource-oriented projects; and

(H) assist the Council on Environmental Quality established by title II of this Act.

42 U.S.C. 4333   SEC. 103. All agencies of the Federal Government shall review their present statutory authority, administrative regulations, and current policies and procedures for the purpose of determining whether there are any deficiencies or inconsistencies therein which prohibit full compliance with the purposes and provisions of this Act and shall propose to the President not later than July 1, 1971, such measures as may be necessary to bring their authority and policies into conformity with the intent, purposes, and procedures set forth in this Act.

42 U.S.C. 4334   SEC. 104. Nothing in Section 102 or 103 shall in any way affect the specific statutory obligations of any Federal

agency (1) to comply with criteria or standards of environmental quality, (2) to coordinate or consult with any other Federal or State agency, or (3) to act, or refrain from acting contingent upon the recommendations or certification of any other Federal or State agency.

SEC. 105. The policies and goals set forth in this Act are supplementary to those set forth in this Act are supplementary to those set forth in existing authorizations of Federal agencies. `42 U.S.C. 4335`

TITLE II

COUNCIL ON ENVIRONMENTAL QUALITY

SEC. 201. The President shall transmit to the Congress annually beginning July 1, 1970, an Environmental Quality Report (hereinafter referred to as the "report") which shall set forth (1) the status and condition of the major natural, manmade, or altered environmental classes of the Nation, including, but not limited to, the air, the aquatic, including marine, estuarine, and fresh water, and the terrestrial environment, including, but not limited to, the forest, dryland, wetland, range, urban, suburban, and rural environment; (2) current and foreseeable trends in the quality, management and utilization of such environments and the effects of those trends on the social, economic, and other requirements of the Nation; (3) the adequacy of available natural resources for fulfilling human and economic requirements of the Nation in the light of expected population pressures; (4) a review of the programs and activities (including regulatory activities) of the Federal Government, the State and local governments, and nongovernmental entities or individuals, with particular reference to their effect on the environment and on the conservation, development and utilization of natural resources; and (5) a program for remedying the deficiencies of existing programs and activities, together with recommendations for legislation. `42 U.S.C. 4341`

SEC. 202. There is created in the Executive Office of the President a Council on Environmental Quality (hereinafter referred to as the "Council"). The Council shall be com- `42 U.S.C. 4342` posed of three members who shall be appointed by the President to serve at his pleasure, by and with the advice and consent of the Senate. The President shall designate one of the members of the Council to serve as Chairman. Each member shall be a person who, as a result of his training, experience, and attainments, is exceptionally well qualified to analyze and interpret environmental trends and information of all kinds; to appraise programs and activities of the Federal Government in the light of the policy set forth in title I of this Act; to be conscious of and responsive to the scientific, economic, social, esthetic, and cultural needs and interests of the Nation; and to formulate and recommend national policies to promote the improvement of the quality of the environment.

42 U.S.C. 4343   SEC. 203. The Council may employ such officers and employees as may be necessary to carry out its functions under this Act. In addition, the Council may employ and fix the compensation of such experts and consultants as may be necessary for the carrying out of its functions under this Act, in accordance with section 3109 of title 5, United States Code (but without regard to the last sentence thereof).

42 U.S.C. 4344   SEC. 204. It shall be the duty and function of the Council—

(1) to assist and advise the President in the preparation of the Environmental Quality Report required by section 201;

(2) to gather timely and authoritative information concerning the conditions and trends in the quality of the environment both current and prospective, to analyze and interpret such information for the purpose of determining whether such conditions and trends are interfering, or are likely to interfere, with the achievement of the policy set forth in title I of this Act, and to compile and submit to the President studies relating to such conditions and trends;

(3) to review and appraise the various programs and activities of the Federal Government in the light of the policy set forth in title I of this Act for the purpose of determining the extent to which such programs and activities are contributing to the achievement of such policy, and to make recommendations to the President with respect thereto;

(4) to develop and recommend to the President national policies to foster and promote the improvement of environmental quality to meet the conservation, social, economic, health, and other requirements and goals of the Nation;

(5) to conduct investigations, studies, surveys, research, and analyses relating to ecological systems and environmental quality;

(6) to document and define changes in the natural environment, including the plant and animal systems, and to accumulate necessary data and other information for a continuing analysis of these changes or trends and an interpretation of their underlying causes;

(7) to report at least once each year to the President on the state and condition of the environment; and

(8) to make and furnish such studies, reports thereon, and recommendations with respect to matters of policy and legislation as the President may request.

SEC. 205. In exercising its powers, functions, and duties under this Act, the Council shall— 42 U.S.C. 4345

(1) consult with the Citizens' Advisory Committee on Environmental Quality established by Executive Order numbered 11472, dated May 29, 1969, and with such representatives of science, industry, agriculture, labor, conservation, organizations, State and local governments and other groups, as it deems advisable; and

(2) utilize, to the fullest extent possible, the services, facilities, and information (including statistical information) of public and private agencies and organizations, and individuals, in order that duplication of effort and expense may be avoided, thus assuring that the Council's activities will not unnecessarily overlap or conflict with similar activities authorized by law and performed by established agencies.

SEC. 206. Members of the Council shall serve full time and the Chairman of the Council shall be compensated at the rate provided for Level II of the Executive Schedule Pay Rates (5 U.S.C. 5313). The other members of the Council shall be compensated at the rate provided for Level IV or the Executive Schedule Pay Rates (5 U.S.C. 5315). 42 U.S.C. 4346

42 U.S.C. 4347   SEC. 207. There are authorized to be appropriated to carry out the provisions of this Act not to exceed $300,000 for fiscal year 1970, $700,000 for fiscal year 1971, and $1,000,000 for each fiscal year thereafter.

LEGISLATIVE HISTORY

HOUSE REPORTS: No. 91–378, pt. 2, accompanying H.R. 12549 (Committee on Merchant Marine and Fisheries) and 91–765 (Committee of Conference).

SENATE REPORT No. 91–296 (Committee on Interior and Insular Affairs).

CONGRESSIONAL RECORD, Vol. 115 (1969):

July 10: Considered and passed Senate.

Sept. 23: Considered and passed House, amended, in lieu of H.R. 12549.

Oct. 8: Senate disagreed to House amendments; agreed to conference.

Dec. 20: Senate agreed to conference report.

Dec. 22: House agreed to conference report.

Appendix B

Council on Environmental Quality Guidelines for Preparation of Environmental Impact Statements

Title 40—Protection of the Environment

CHAPTER V—COUNCIL ON ENVIRONMENTAL QUALITY

PART 1500—PREPARATION OF ENVIRONMENTAL IMPACT
STATEMENTS: GUIDELINES

On May 2, 1973, the Council on Environmental Quality published in the
FEDERAL REGISTER, for public comment, a proposed revision of its
guidelines for the preparation of environmental impact statements.

Pursuant to the National Environmental Policy Act (P.L. 91–190, 42 U.S.C. 4321 et seq.) and Executive Order 11514 (35 FR 4247) all Federal departments, agencies, and establishments are required to prepare such statements in connection with their proposals for legislation and other major Federal actions significantly affecting the quality of the human environment. The authority for the Council's guidelines is set forth below in § 1500.1. The specific policies to be implemented by the guidelines is set forth below in § 1500.2.

The Council received numerous comments on its proposed guidelines from environmental groups, Federal, State, and local agencies, industry, and private individuals. Two general themes were presented in the majority of the comments. First, the Council should increase the opportunity for public involvement in the impact statement process. Second, the Council should provide more detailed guidance on the responsibilities of Federal agencies in light of recent court decisions interpreting the Act. The proposed guidelines have been revised in light of the specific comments relating to these general themes, as well as other comments received, and are now being issued in final form.

The guidelines will appear in the Code of Federal Regulations in Title 40, Chapter V, at Part 1500. They are being codified, in part, because they affect State and local governmental agencies, environmental groups, industry, and private individuals, in addition to Federal agencies, to which they are specifically directed, and the resultant need to make them widely and readily available.

Sec.

1500.13   Application of section 102(2)(C) procedure to existing projects and programs.

1500.14   Supplementary guidelines; evaluation of procedures.

Sec.

Appendix I Summary to accompany draft and final statements.

Appendix II Areas of environmental impact and Federal agencies and Federal State agencies with jurisdiction by law or special expertise to comment thereon.

Appendix III Offices within Federal agencies and Federal-State agencies for information regarding the agencies' NEPA activities and for receiving other agencies' impact statements for which comments are requested.

Appendix IV State and local agency review of impact statements.

AUTHORITY: National Environmental Act (P.L. 91–190, 42 U.S.C. 4321 et seq.) and Executive Order 11514.

† **1500.1   Purpose and authority.**

(a) This directive provides guidelines to Federal departments, agencies, and establishments for preparing detailed environmental statements on proposals for legislation and other major Federal actions significantly affecting the quality of the human environment as required by section 102(2)(C) of the National Environmental Policy Act (P.L. 91–190, 42 U.S.C. 4321 et. seq.) (hereafter "the Act"). Underlying the preparation of such environmental statements is the mandate of both the Act and Executive Order 11514 (35 FR 4247) of March 5, 1970, that all Federal agencies, to the fullest extent possible, direct their policies, plans and programs to protect and enhance environmental quality. Agencies are required to view their actions in a manner calculated to encourage productive and enjoyable harmony between man and his environment, to promote efforts preventing or eliminating damage to the environment and biosphere and stimulating the health and welfare of man, and to enrich the understanding of the ecological systems and natural resources important to the Nation. The objective of section 102(2)(C) of the Act and of these guidelines is to assist agencies in implementing these policies. This requires agencies to build into their decisionmaking process, beginning at the earliest possible point, an appropriate and careful consideration of the environmental aspects of proposed action in order that adverse environmental effects may be avoided or minimized and environmental quality previously lost may be restored. This directive also provides guidance to Federal, State, and local agencies and the public in commenting on statements prepared under these guidelines.

(b) Pursuant to section 204(3) of the Act the Council on Environmental Quality (hereafter "the Council") is assigned the duty and function of reviewing and appraising the programs and activities of the Federal Government, in the light of the Act's policy, for the purpose of determining the extent to which such programs and activities are contributing to the achievement of such policy, and to make recommendations to the President with respect thereto. Section 102(2)(B) of the Act directs all Federal agencies to identify and develop methods and procedures, in consultation with the Council, to insure that unquantified environmental values be given appropriate consideration in decisionmaking along with economic and technical considerations; section 102(2)(C) of the Act directs that copies of all environmental impact statements be filed with the Council; and section 102(2)(H) directs all Federal agencies to assist the Council in the performance of its functions. These provisions have been supplemented in sections 3(h) and (i) of Executive Order 11514 by directions that the Council issue guidelines to Federal agencies for preparation of environmental impact statements and such other instructions to agencies and requests for reports and information as may be required to carry out the Council's responsibilities under the Act.

§ 1500.2 Policy.

(a) As early as possible and in all cases prior to agency decision concerning recommendations or favorable reports on proposals for (1) legislation significantly affecting the quality of the human environment (see §§ 1500.5(i) and 1500.12) (hereafter "legislative actions") and (2) all other major Federal actions significantly affecting the quality of the human environment (hereafter "administrative actions"), Federal agencies will, in consultation with other appropriate Federal, State and local agencies and the public assess in detail the potential environmental impact.

(b) Initial assessments of the environmental impacts of proposed action should be undertaken concurrently with initial technical and economic studies and, where required, a draft environmental impact statement prepared and circulated for comment in time to accompany the proposal through the existing agency review processes for such action. In this process, Federal agencies shall: (1) Provide for circulation of draft environmental statements to other Federal, State, and local agencies and for their availability to the public in accordance with the provisions of these guidelines; (2) consider the comments of the agencies and the public; and (3) issue final environmental impact statements responsive to the comments received. The purpose of this assessment and consultation process is to provide agencies and other decisionmakers as well as members of the public with an understanding of the potential environmental effects of proposed actions, to avoid or minimize adverse effects wherever possible, and to restore or enhance environmental quality to the fullest extent practicable. In particular, agencies should use the environmental impact statement process to explore alternative actions that will avoid or minimize adverse impacts and to evaluate both the long- and short-range implications of proposed actions to man, his physical and social surroundings, and to nature. Agencies should consider the results of their environmental assessments along with their assessments of the net economic, technical and other benefits of proposed actions and use all practicable means, consistent with other essential considerations of national policy, to restore environmental quality as well as to avoid or minimize undesirable consequences for the environment.

§ 1500.3   Agency and OMB procedures.

(a) Pursuant to section 2(f) of Executive Order 11514, the heads of Federal agencies have been directed to proceed with measures required by section 102(2)(C) of the Act. Previous guidelines of the Council directed each agency to establish its own formal procedures for (1) identifying those agency actions requiring environmental statements, the appropriate time prior to decision for the consultations required by section 102(2)(C) and the agency review process for which environmental statements are to be available, (2) obtaining information required in their preparation, (3) designating the officials who are to be responsible for the statements, (4) consulting with and taking account of the comments of appropriate Federal, State and local agencies and the public, including obtaining the comment of the Administrator of the Environmental Protection Agency when required under section 309 of the Clean Air Act, as amended, and (5) meeting the requirements of section 2(b) of Executive Order 11514 for providing timely public information on Federal plans and programs with environmental impact. Each agency, including both departmental and subdepartmental components having such procedures, shall review its procedures and shall revise them, in consultation with the Council, as may be necessary in order to respond to requirements imposed by these revised guidelines as well as by such previous directives. After such consultation, proposed revisions of such agency procedures shall be published in the FEDERAL REGISTER no later than October 30, 1973. A minimum 45-day period for public comment shall be provided, followed by publication of final procedures no later than forty-five (45) days after the conclusion of the comment period. Each agency shall submit seven (7) copies of all such procedures to the Council. Any future revision of such agency procedures shall similarly be proposed and adopted only after prior consultation with the Council and, in the case of substantial revision, opportunity for public comment. All revisions shall be published in the FEDERAL REGISTER.

(b) Each Federal agency should consult, with the assistance of the Council and the Office of Management and Budget if desired, with other appropriate Federal agencies in the development and revision of the above procedures so as to achieve consistency in dealing with similar activities and to assure effective coordination among agencies in their review of proposed activities. Where applicable, State and local review of such agency procedures should be conducted pursuant to procedures established by Office of Management and Budget Circular No. A-85.

(c) Existing mechanisms for obtaining the views of Federal, State, and local agencies on proposed Federal actions should be utilized to the maximum extent practicable in dealing with environmental matters. The Office of Management and Budget will issue instructions, as necessary, to take full advantage of such existing mechanisms.

§ 1500.4   Federal agencies included; effect of the Act on existing agency mandates.

(a) Section 102(2)(C) of the Act applies to all agencies of the Federal Government. Section 102 of the Act provides that "to the fullest extent possible: (1) The policies, regulations, and public laws of the United States shall be interpreted and administered in accordance with the policies set forth in this Act," and section 105 of the Act provides that "the policies and goals set forth in this Act are supplementary to those set forth in existing authorizations of Federal agencies." This means that each agency shall interpret the provisions of the Act as a supplement to its existing authority and as a mandate to view traditional policies and missions in the light of the Act's national environmental objectives. In accordance with this purpose, agencies should continue to review their policies, procedures, and regulations and to revise them as necessary to ensure full compliance with the purposes and provisions of the Act. The phrase "to the fullest extent possible" in section 102 is meant to make clear that each agency of the Federal Government shall comply with that section unless existing law applicable to the agency's operations expressly prohibits or makes compliance impossible.

§ 1500.5   Types of actions covered by the Act.

(a) "Actions" include but are not limited to:

(1) Recommendations or favorable reports relating to legislation in-

cluding requests for appropriations. The requirement for following the section 102(2)(C) procedure as elaborated in these guidelines applies to both (i) agency recommendations on their own proposals for legislation (see § 1500.12); and (ii) agency reports on legislation initiated elsewhere. In the latter case only the agency which has primary responsibility for the subject matter involved will prepare an environmental statement.

(2) New and continuing projects and program activities: directly undertaken by Federal agencies: or supported in whole or in part through Federal contracts, grants, subsidies, loans, or other forms of funding assistance (except where such assistance is solely in the form of general revenue sharing funds, distributed under the State and Local Fiscal Assistance Act of 1972, 31 U.S.C. 1221 et. seq. with no Federal agency control over the subsequent use of such funds); or involving a Federal lease, permit, license certificate or other entitlement for use.

(3) The making, modification, or establishment of regulations, rules, procedures, and policy.

§ 1500.6   Identifying major actions significantly affecting the environment.

(a) The statutory clause "major Federal actions significantly affecting the quality of the human environment" is to be construed by agencies with a view to the overall, cumulative impact of the action proposed, related Federal actions and projects in the area, and further actions contemplated. Such actions may be localized in their impact, but if there is potential that the environment may be significantly affected, the statement is to be prepared. Proposed major actions, the environmental impact of which is likely to be highly controversial, should be covered in all cases. In considering what constitutes major action significantly affecting the environment, agencies should bear in mind that the effect of many Federal decisions about a project or complex of projects can be individually limited but cumulatively considerable. This can occur when one or more agencies over a period of years puts into a project individually minor but collectively major resources, when one decision involving a limited amount of money is a precedent for action in much larger cases or represents a decision in principle about a future major course of action, or when several Government agencies individually make decisions about partial aspects of a major action. In all such cases, an environmental statement should be prepared if it is reasonable to anticipate a cumulatively significant impact on the environment from Federal action. The Council, on the basis of a written assessment of the impacts involved, is available to assist agencies in determining whether specific actions require impact statements.

(b) Section 101(b) of the Act indicates the broad range of aspects of the environment to be surveyed in any assessment of significant effect. The Act also indicates that adverse significant effects include those that degrade the quality of the environment, curtail the range of beneficial uses of the environment, and serve short-term, to the disadvantage of long-term, environmental goals. Significant effects can also include actions which may have both beneficial and detrimental effects, even if on balance the agency believes that the effect will be beneficial. Significant effects also include secondary effects, as described more fully, for example, in § 1500.8(a)(iii)(B). The significance of a proposed action may also vary with the setting, with the result that an action that would have little impact in an urban area may be significant in a rural setting or vice versa. While a precise definition of environmental "significance," valid in all contexts, is not possible, effects to be considered in assessing significance include, but are not limited to, those outlined in Appendix II of these guidelines.

(c) Each of the provisions of the Act, except section 102(2)(C), applies to all Federal agency actions. Section 102(2)(C) requires the preparation of a detailed environmental impact statement in the case of "major Federal actions significantly affecting the quality of the human environment." The identification of major actions significantly affecting the environment is the responsibility of each Federal agency, to be carried out against the background of its own particular operations. The action must be a (1) "major" action, (2) which is a "Federal action," (3) which has a "significant" effect, and (4) which involves the "quality of the human environment." The words "major" and "significantly" are intended to imply thresholds of importance and impact that must be met before a statement is required. The action causing the impact must also be one where there is sufficient Federal control and responsibility to constitute "Federal action" in contrast to cases where such Federal control and responsibility are not present as, for example, when Federal funds are distributed in the form of general revenue sharing to be used by State and local governments (see § 1500.5(ii)). Finally, the action must be one that significantly affects the quality of the human environment either by directly affecting human beings or by indirectly affecting human beings through adverse effects on the environment. Each agency should review the typical classes of actions that it undertakes and, in consultation with the Council, should develop specific criteria and methods for identifying those actions likely to require environmental statements and those actions likely not to require environmental statements. Normally this will involve:

(i) Making an initial assessment of the environmental impacts typically associated with principal types of agency action.

(ii) Identifying on the basis of this assessment, types of actions which normally do, and types of actions which normally do not, require statements.

(iii) With respect to remaining actions that may require statements depending on the circumstances, and those actions determined under the preceding paragraph (C)(4)(ii) of this section as likely to require statements, identifying: (*a*) what basic information needs to be gathered; (*b*) how and when such information is to be assembled and analyzed; and (*c*) on what basis environmental assessments and decisions to prepare impact statements will be made. Agencies may either include this substantive guidance in the procedures issued pursuant to § 1500.3(a) of these guidelines, or issue such guidance as supplemental instructions to aid relevant agency personnel in implementing the impact statement process. Pursuant to § 1500.14 of these guidelines, agencies shall report to the Council by June 30, 1974, on the progress made in developing such substantive guidance.

(d) (1) Agencies should give careful attention to identifying and defining the purpose and scope of the action which would most appropriately serve as the subject of the statement. In many cases, broad program statements will be required in order to assess the environmental effects of a number of individual actions on a given geographical area (e.g., coal leases), or environmental impacts that are generic or common to a series of agency actions (e.g., maintenance or waste handling practices), or the overall impact of a large-scale program or chain of contemplated projects (e.g., major lengths of highway as opposed to small segments). Subsequent statements on major individual actions will be necessary where such actions have significant environmental impacts not adequately evaluated in the program statement.

(2) Agencies engaging in major technology research and development programs should develop procedures for periodic evaluation to determine when a program statement is required for such programs. Factors to be considered in making this determination include the magnitude of Federal investment in the program, the likelihood of widespread application of the technology, the degree of environmental impact which would occur if the technology were widely applied, and the extent to which continued investment in the new technology is likely to restrict future alternatives. Statements must be written late enough in the development process to contain meaningful information, but early enough so that this information can practically serve as an input in the decision-making process. Where it is anticipated that a statement may ultimately be required but that its preparation is still premature, the agency should prepare an evaluation briefly setting forth the reasons for its determination that a statement is

not yet necessary. This evaluation should be periodically updated, particularly when significant new information becomes available concerning the potential environmental impact of the program. In any case, a statement must be prepared before research activities have reached a stage of investment or commitment to implementation likely to determine subsequent development or restrict later alternatives. Statements on technology research and development programs should include an analysis not only of alternative forms of the same technology that might reduce any adverse environmental impacts but also of alternative technologies that would serve the same function as the technology under consideration. Efforts should be made to involve other Federal agencies and interested groups with relevant expertise in the preparation of such statements because the impacts and alternatives to be considered are likely to be less well defined than in other types of statements.

(e) In accordance with the policy of the Act and Executive Order 11514 agencies have a responsibility to develop procedures to insure the fullest practicable provision of timely public information and understanding of Federal plans and programs with environmental impact in order to obtain the views of interested parties. In furtherance of this policy, agency procedures should include an appropriate early notice system for informing the public of the decision to prepare a draft environmental statement on proposed administrative actions (and for soliciting comments that may be helpful in preparing the statement) as soon as is practicable after the decision to prepare the statement is made. In this connection, agencies should: (1) maintain a list of administrative actions for which environmental statements are being prepared; (2) revise the list at regular intervals specified in the agency's procedures developed pursuant to § 1500.3(a) of these guidelines (but not less than quarterly) and transmit each such revision to the Council; and (3) make the list available for public inspection on request. The Council will periodically publish such lists in the FEDERAL REGISTER. If an agency decides that an environmental statement is not necessary for a proposed action (i) which the agency has identified pursuant to § 1500.6(c)(4)(ii) as normally requiring preparation of a statement, (ii) which is similar to actions for which the agency has prepared a significant number of statements, (iii) which the agency has previously announced would be the subject of a statement, or (iv) for which the agency has made a negative determination in response to a request from the Council pursuant to § 1500.11(f), the agency shall prepare a publicly available record briefly setting forth the agency's decision and the reasons for that determination. Lists of such negative determinations, and any evaluations made pursuant to § 1500.6 which conclude that preparation of a statement is not yet timely, shall be prepared and made available in the same manner as provided in this subsection for lists of statements under preparation.

§ 1500.7   Preparing draft environmental statements; public hearings.

(a) Each environmental impact statement shall be prepared and circulated in draft form for comment in accordance with the provisions of these guidelines. The draft statement must fulfill and satisfy to the fullest extent possible at the time the draft is prepared the requirements established for final statements by section 102(2)(C). (Where an agency has an established practice of declining to favor an alternative until public comments on a proposed action have been received, the draft environmental statement may indicate that two or more alternatives are under consideration.) Comments received shall be carefully evaluated and considered in the decision process. A final statement with substantive comments attached shall then be issued and circulated in accordance with applicable provisions of §§ 1500.10, 1500.11, or 1500.12. It is important that draft environmental statements be prepared and circulated for comment and furnished to the Council as early as possible in the agency review process in order to permit agency decisionmakers and outside reviewers to give meaningful consideration to the environmental issues involved. In particular, agencies should keep in mind that such statements are to serve as the means of assessing the environmental impact of proposed agency actions, rather than as a justification for decisions already made. This means that draft statements on administrative actions should be prepared and circulated for comment prior to the first significant point of decision in the agency review process. For major categories of agency action, this point should be identified in the procedures issued pursuant to § 1500.3(a). For major categories of projects involving an applicant and identified pursuant to § 1500.6(*c*)(c)(ii) as normally requiring the preparation of a statement, agencies should include in their procedures provisions limiting actions which an applicant is permitted to take prior to completion and review of the final statement with respect to his application.

(b) Where more than one agency (1) directly sponsors an action, or is directly involved in an action through funding, licenses, or permits, or (2) is involved in a group of actions directly related to each other because of their functional interdependence and geographical proximity, consideration should be given to preparing one statement for all the Federal actions involved (see § 1500.6(d)(1)). Agencies in such cases should consider the possibility of joint preparation of a statement by all agencies concerned, or designation of a single "lead agency" to assume supervisory responsibility for preparation of the statement. Where a lead agency prepares the statement, the other agencies involved should provide assistance with respect to their areas of jurisdiction and expertise. In either case, the statement should contain an environmental assessment of the full range of Federal actions involved, should reflect the views of all participating agencies, and should be prepared before major or irreversible actions have been taken by any of the participating agencies. Factors relevant in determining an appropriate lead agency include the time sequence in which the agencies become involved, the magnitude of their respective involvement, and their relative expertise with respect to the project's environmental effects. As necessary, the Council will assist in resolving questions of responsibility for statement preparation in the case of multiagency actions. Federal Regional Councils, agencies and the public are encouraged to bring to the attention of the Council and other relevant agencies appropriate situations where a geographic or regionally focused statement would be desirable because of the cumulative environmental effects likely to result from multi-agency actions in the area.

(c) Where an agency relies on an applicant to submit initial environmental information, the agency should assist the applicant by outlining the types of information required. In all cases, the agency should make its own evaluation of the environmental issues and take responsibility for the scope and content of draft and final environmental statements.

(d) Agency procedures developed pursuant to § 1500.3(a) of these guidelines should indicate as explicitly as possible those types of agency decisions or actions which utilize hearings as part of the normal agency review process, either as a result of statutory requirement or agency practice. To the fullest extent possible, all such hearings shall include consideration of the environmental aspects of the proposed action. Agency procedures shall also specifically include provision for public hearings on major actions with environmental impact, whenever appropriate, and for providing the public with relevant information, including information on alternative courses of action. In deciding whether a public hearing is appropriate, an agency should consider: (1) The magnitude of the proposal in terms of economic costs, the geographic area involved, and the uniqueness or size of commitment of the resources involved; (2) the degree of interest in the proposal, as evidenced by requests from the public and from Federal, State and local authorities that a hearing be held; (3) the complexity of the issue and the likelihood that information will be presented at the hearing which will be of assistance to the agency in fulfilling its responsibilities under the Act; and (4) the extent to which public involvement already has been achieved through other means, such as earlier public hearings, meetings with citizen representatives, and/or written comments on the proposed action. Agencies should make any draft environmental statements to be issued available to the public at least fifteen (15) days prior to the time of such hearings.

§ 1500.8 Content of environmental statements.

(a) The following points are to be covered:

(1) A description of the proposed action, a statement of its purposes, and a description of the environment affected, including information, summary technical data, and maps and diagrams where relevant, adequate to permit an assessment of potential environmental impact by commenting agencies and the public. Highly technical and specialized analyses and data should be avoided in the body of the draft impact statement. Such materials should be attached as appendices or footnoted with adequate bibliographic references. The statement should also succinctly describe the environment of the area affected as it exists prior to a proposed action, including other Federal activities in the area affected by the proposed action which are related to the proposed action. The interrelationships and cumulative environmental impacts of the proposed action and other related Federal projects shall be presented in the statement. The amount of detail provided in such descriptions should be commensurate with the extent and expected impact of the action, and with the amount of information required at the particular level of decisionmaking (planning, feasibility, design, etc.). In order to ensure accurate descriptions and environmental assessments, site visits should be made where feasible. Agencies should also take care to identify, as appropriate, population and growth characteristics of the affected area and any population and growth assumptions used to justify the project or program or to determine secondary population and growth impacts resulting from the proposed action and its alternatives (see paragraph (a)(1)(3)(ii), of this section). In discussing these population aspects, agencies should give consideration to using the rates of growth in the region of the project contained in the projection compiled for the Water Resources Council by the Bureau of Economic Analysis of the Department of Commerce and the Economic Research Service of the Department of Agriculture (the "OBERS" projection). In any event it is essential that the sources of data used to identify, quantify or evaluate any and all environmental consequences be expressly noted.

(2) The relationship of the proposed action to land use plans, policies, and controls for the affected area. This requires a discussion of how the proposed action may conform or conflict with the objectives and specific

terms of approved or proposed Federal, State, and local land use plans, policies, and controls, if any, for the area affected including those developed in response to the Clean Air Act or the Federal Water Pollution Control Act Amendments of 1972. Where a conflict or inconsistency exists, the statement should describe the extent to which the agency has reconciled its proposed action with the plan, policy or control, and the reasons why the agency has decided to proceed notwithstanding the absence of full reconciliation.

(3) The probable impact of the proposed action on the environment.

(i) This requires agencies to assess the positive and negative effects of the proposed action as it affects both the national and international environment. The attention given to different environmental factors will vary according to the nature, scale, and location of proposed actions. Among factors to consider should be the potential effect of the action on such aspects of the environment as those listed in Appendix II of these guidelines. Primary attention should be given in the statement to discussing those factors most evidently impacted by the proposed action.

(ii) Secondary or indirect, as well as primary or direct, consequences for the environment should be included in the analysis. Many major Federal actions, in particular, those that involve the construction or licensing of infrastructure investments (e.g., highways, airports, sewer systems, water resource projects, etc.), stimulate or induce secondary effects in the form of associated investments and changed patterns of social and economic activities. Such secondary effects, through their impacts on existing community facilities and activities, through inducing new facilities and activities, or through changes in natural conditions, may often be even more substantial than the primary effects of the original action itself. For example, the effects of the proposed action on population and growth may be among the more significant secondary effects. Such population and growth impacts should be estimated if expected to be significant (using data identified as indicated in § 1500.8(a)(1) and an assessment made of the effect of any possible change in population patterns or growth upon the resource base, including land use, water, and public services, of the area in question.

(4) Alternatives to the proposed action, including, where relevant, those not within the existing authority of the responsible agency. (Section 102(2)(D) of the Act requires the responsible agency to "study, develop, and describe appropriate alternatives to recommended courses of action in any proposal which involves unresolved conflicts concerning alternative uses of available resources"). A rigorous exploration and objective evaluation of the environmental impacts of all reasonable alternative ac-

tions, particularly those that might enhance environmental quality or avoid some or all of the adverse environmental effects, is essential. Sufficient analysis of such alternatives and their environmental benefits, costs and risks should accompany the proposed action through the agency review process in order not to foreclose prematurely options which might enhance environmental quality or have less detrimental effects. Examples of such alternatives include: the alternative of taking no action or of postponing action pending further study; alternatives requiring actions of a significantly different nature which would provide similar benefits with different environmental impacts (e.g., nonstructural alternatives to flood control programs, or mass transit alternatives to highway construction); alternatives related to different designs or details of the proposed action which would present different environmental impacts (e.g., cooling ponds vs. cooling towers for a power plant or alternatives that will significantly conserve energy); alternative measures to provide for compensation of fish and wildlife losses, including the acquisition of land, waters, and interests therein. In each case, the analysis should be sufficiently detailed to reveal the agency's comparative evaluation of the environmental benefits, costs and risks of the proposed action and each reasonable alternative. Where an existing impact statement already contains such an analysis, its treatment of alternatives may be incorporated provided that such treatment is current and relevant to the precise purpose of the proposed action.

(5) Any probable adverse environmental effects which cannot be avoided (such as water or air pollution, undesirable land use patterns, damage to life systems, urban congestion, threats to health or other consequences adverse to the environmental goals set out in section 101(b) of the Act). This should be a brief section summarizing in one place those effects discussed in paragraph (a)(3) of this section that are adverse and unavoidable under the proposed action. Included for purposes of contrast should be a clear statement of how other avoidable adverse effects discussed in paragraph (a)(2) of this section will be mitigated.

(6) The relationship between local short-term uses of man's environment and the maintenance and enhancement of long-term productivity. This section should contain a brief discussion of the extent to which the proposed action involves tradeoffs between short-term environmental gains at the expense of long-term losses, or vice versa, and a discussion of the extent to which the proposed action forecloses future options. In this context short-term and long-term do not refer to any fixed time periods, but should be viewed in terms of the environmentally significant consequences of the proposed action.

(7) Any irreversible and irretrievable commitments of resources that would be involved in the proposed action should it be implemented. This requires the agency to identify from its survey of unavoidable impacts in paragraph (a)(5) of this section the extent to which the action irreversibly curtails the range of potential uses of the environment. Agencies should avoid construing the term "resources" to mean only the labor and materials devoted to an action. "Resources" also means the natural and cultural resources committed to loss or destruction by the action.

(8) An indication of what other interests and considerations of Federal policy are thought to offset the adverse environmental effects of the proposed action identified pursuant to paragraphs (a)(3) and (5) of this section. The statement should also indicate the extent to which these stated countervailing benefits could be realized by following reasonable alternatives to the proposed action (as identified in paragraph (a)(4) of this section) that would avoid some or all of the adverse environmental effects. In this connection, agencies that prepare cost-benefit analyses of proposed actions should attach such analyses, or summaries thereof, to the environmental impact statement, and should clearly indicate the extent to which environmental costs have not been reflected in such analyses.

(b) In developing the above points agencies should make every effort to convey the required information succinctly in a form easily understood, both by members of the public and by public decisionmakers, giving attention to the substance of the information conveyed rather than to the particular form, or length, or detail of the statement. Each of the above points, for example, need not always occupy a distinct section of the statement if it is otherwise adequately covered in discussing the impact of the proposed action and its alternatives—which items should normally be the focus of the statement. Draft statements should indicate at appropriate points in the text any underlying studies, reports, and other information obtained and considered by the agency in preparing the statement including any cost-benefit analyses prepared by the agency, and reports of consulting agencies under the Fish and Wildlife Coordination Act, 16 U.S.C. 661 et seq., and the National Historic Preservation Act of 1966, 16 U.S.C. 470 et seq., where such consultation has taken place. In the case of documents not likely to be easily accessible (such as internal studies or reports), the agency should indicate how such information may be obtained. If such information is attached to the statement, care should be taken to ensure that the statement remains an essentially self-contained instrument, capable of being understood by the reader without the need for undue cross reference.

(c) Each environmental statement should be prepared in accordance with the precept in section 102(2)(A) of the Act that all agencies of the Federal Government "utilize a systematic, interdisciplinary approach which will insure the integrated use of the natural and social sciences and the environmental design arts in planning and decisionmaking which may have an impact on man's environment." Agencies should attempt to have relevant disciplines represented on their own staffs; where this is not feasible they should make appropriate use of relevant Federal, State, and local agencies or the professional services of universities and outside consultants. The interdisciplinary approach should not be limited to the preparation of the environmental impact statement, but should also be used in the early planning stages of the proposed action. Early application of such an approach should help assure a systematic evaluation of reasonable alternative courses of action and their potential social, economic, and environmental consequences.

(d) Appendix I prescribes the form of the summary sheet which should accompany each draft and final environmental statement.

§ 1500.9   Review of draft environmental statements by Federal, Federal-State, State, and local agencies and by the public.

(a) *Federal agency review.* (1) *In general.* A Federal agency considering an action requiring an environmental statement should consult with, and (on the basis of a draft environmental statement for which the agency takes responsibility) obtain the comment on the environmental impact of the action of Federal and Federal-State agencies with jurisdiction by law or special expertise with respect to any environmental impact involved. These Federal and Federal-State agencies and their relevant areas of expertise include those indentified in Appendices II and III to these guidelines. It is recommended that the listed departments and agencies establish contact points, which may be regional offices, for providing comments on the environmental statements. The requirement in section 102(2)(C) to obtain comment from Federal agencies having jurisdiction or special expertise is in addition to any specific statutory obligation of any Federal agency to coordinate or consult with any other Federal or State agency. Agencies should, for example, be alert to consultation requirements of the Fish and Wildlife Coordination Act, 16 U.S.C. 661 et seq., and the National Historic Preservation Act of 1966, 16 U.S.C. 470 et seq. To the extent possible, statements or findings concerning environmental impact required by other statutes, such as section 4(f) of the Department of Transportation Act of 1966, 49 U.S.C. 1653(f), or section 106 of the National Historic Preservation Act of 1966, should be combined with compliance with the environmental impact statement requirements of section

102(2)(C) of the Act to yield a single document which meets all applicable requirements. The Advisory Council on Historic Preservation, the Department of Transportation, and the Department of the Interior, in consultation with the Council, will issue any necessary supplementing instructions for furnishing information or findings not forthcoming under the environmental impact statement process.

(b) *EPA review.* Section 309 of the Clean Air Act, as amended (42 U.S.C. § 1857h-7), provides that the Administrator of the Environmental Protection Agency shall comment in writing on the environmental impact of any matter relating to his duties and responsibilities, and shall refer to the Council any matter that the Administrator determines is unsatisfactory from the standpoint of public health or welfare or environmental quality. Accordingly, wherever an agency action related to air or water quality, noise abatement and control, pesticide regulation, solid waste disposal, generally applicable environmental radiation criteria and standards, or other provision of the authority of the Administrator is involved, Federal agencies are required to submit such proposed actions and their environmental impact statements, if such have been prepared, to the Administrator for review and comment in writing. In all cases where EPA determines that proposed agency action is environmentally unsatisfactory, or where EPA determines that an environmental statement is so inadequate that such a determination cannot be made, EPA shall publish its determination and notify the Council as soon as practicable. The Administrator's comments shall constitute his comments for the purposes of both section 309 of the Clean Air Act and section 102(2)(C) of the National Environmental Policy Act.

(c) State and local review. Office of Management and Budget Circular No. A-95 (Revised) through its system of State and areawide clearinghouses provides a means for securing the views of State and local environmental agencies, which can assist in the preparation and review of environmental impact statements. Current instructions for obtaining the views of such agencies are contained in the joint OMB-CEQ memorandum attached to these guidelines as Appendix IV. A current listing of clearinghouses is issued periodically by the Office of Management and Budget.

(d) *Public review.* The procedures established by these guidelines are designed to encourage public participation in the impact statement process at the earliest possible time. Agency procedures should make provision for facilitating the comment of public and private organizations and individuals by announcing the availability of draft environmental statements and by making copies available to organizations and individuals that request an opportunity to comment. Agencies should devise methods for publicizing the existence of draft statements, for example, by publication of notices in local newspapers or by maintaining a list of groups, including relevant conservation commissions, known to be interested in the agency's activities and directly notifying such groups of the existence of a draft statement, or sending them a copy, as soon as it has been prepared. A copy of the draft statement should in all cases be sent to any applicant whose project is the subject of the statement. Materials to be made available to the public shall be provided without charge to the extent practicable, or at a fee which is not more than the actual cost of reproducing copies required to be sent to other Federal agencies, including the Council.

(e) *Responsibilities of commenting entities.* (1) Agencies and members of the public submitting comments on proposed actions on the basis of draft environmental statements should endeavor to make their comments as specific, substantive, and factual as possible without undue attention to matters of form in the impact statements. Although the comments need not conform to any particular format, it would assist agencies reviewing comments if the comments were organized in a manner consistent with the structure of the draft statement. Emphasis should be placed on the assessment of the environmental impacts of the proposed action, and the acceptability of those impacts on the quality of the environment, particularly as contrasted with the impacts of reasonable alternatives to the action. Commenting entities may recommend modifications to the proposed action and/or new alternatives that will enhance environmental quality and avoid or minimize adverse environmental impacts.

(2) Commenting agencies should indicate whether any of their projects not identified in the draft statement are sufficiently advanced in planning and related environmentally to the proposed action so that a discussion of the environmental interrelationships should be included in the final statement (see § 1500.8(a)(1)). The Council is available to assist agencies in making such determinations.

(3) Agencies and members of the public should indicate in their comments the nature of any monitoring of the environmental effects of the proposed project that appears particularly appropriate. Such monitoring may be necessary during the construction, startup, or operation phases of the project. Agencies with special expertise with respect to the environmental impacts involved are encouraged to assist the sponsoring agency in the establishment and operation of appropriate environmental monitoring.

(f) Agencies seeking comment shall establish time limits of not less than forty-five (45) days for reply, after which it may be presumed, unless the agency or party consulted requests a specified extension of time, that the agency or party consulted has no comment to make. Agencies seeking comment should endeavor to comply with requests for extensions of time of up to fifteen (15) days. In determining an appropriate period for comment, agencies should consider the magnitude and complexity of the statement and the extent of citizen interest in the proposed action.

§ 1500.10   Preparation and circulation of final environmental statements.

(a) Agencies should make every effort to discover and discuss all major points of view on the environmental effects of the proposed action and its alternatives in the draft statement itself. However, where opposing professional views and responsible opinion have been overlooked in the draft statement and are brought to the agency's attention through the commenting process, the agency should review the environmental effects of the action in light of those views and should make a meaningful reference in the final statement to the existence of any responsible opposing view not adequately discussed in the draft statement, indicating the agency's response to the issues raised. All substantive comments received on the draft (or summaries thereof where response has been exceptionally voluminous) should be attached to the final statement, whether or not each such comment is thought to merit individual discussion by the agency in the text of the statement.

(b) Copies of final statements, with comments attached, shall be sent to all Federal, State, and local agencies and private organizations that made substantive comments on the draft statement and to individuals who requested a copy of the final statement, as well as any applicant whose project is the subject of the statement. Copies of final statements shall in all cases be sent to the Environmental Protection Agency to assist it in carrying out its responsibilities under section 309 of the Clean Air Act. Where the number of comments on a draft statement is such that distribution of the final statement to all commenting entities appears impracticable, the agency shall consult with the Council concerning alternative arrangements for distribution of the statement.

§ 1500.11   Transmittal of statements to the Council; minimum periods for review; requests by the Council.

(a) As soon as they have been prepared, ten (10) copies of draft environmental statements, five (5) copies of all comments made thereon (to be forwarded to the Council by the entity making comment at the time comment is forwarded to the responsible agency), and ten (10) copies of the final text of environmental statements (together with the substance of all comments received by the responsible agency from Federal, State, and local agencies and from private organizations and individuals) shall be supplied to the Council. This will serve to meet the statutory requirement to make environmental statements available to the President. At the same time that copies of draft and final statements are sent to the Council, copies should also be sent to relevant commenting entities as set forth in §§ 1500.9 and 1500.10(b) of these guidelines.

(b) To the maximum extent practicable no administrative action subject to section 102(2)(C) is to be taken sooner than ninety (90) days after a draft environmental statement has been circulated for comment, furnished to the Council and, except where advance public disclosure will result in significantly increased costs of procurement to the Government, made available to the public pursuant to these guidelines; neither should such administrative action be taken sooner than thirty (30) days after the final text of an environmental statement (together with comments) has been made available to the Council, commenting agencies, and the public. In all cases, agencies should allot a sufficient review period for the final statement so as to comply with the statutory requirement that the "statement and the comments and views of appropriate Federal, State, and local agencies * * * accompany the proposal through the existing agency review process." If the final text of an environmental statement is filed within ninety (90) days after a draft statement has been circulated for comment, furnished to the Council and made public pursuant to this section of these guidelines, the minimum thirty (30) day period and the ninety (90) day period may run concurrently to the extent that they overlap. An agency may at any time supplement or amend a draft or final environmental statement, particularly when substantial changes are made in the proposed action, or significant new information becomes available concerning its environmental aspects. In such cases the agency should consult with the Council with respect to the possible need for or desirability of recirculation of the statement for the appropriate period.

(c) The Council will publish weekly in the FEDERAL REGISTER lists of environmental statements received during the preceding week that are available for public comment. The date of publication of such lists shall be the date from which the minimum periods for review and advance availability of statements shall be calculated.

(d) The Council's publication of notice of the availability of statements is in addition to the agency's responsibility, as described in § 1500.9(d) of these guidelines, to insure the fullest practicable provision of timely public information concerning the existence and availability of environmental statements. The agency responsible for the environmental statement is also responsible for making the statement, the comments received, and any underlying documents available to the public pursuant to the provisions of the Freedom of Information Act (5 U.S.C., 552), without regard to the exclusion of intra- or interagency memoranda when such memoranda transmit comments of Federal agencies on the environmental impact of the proposed action pursuant to § 1500.9 of these guidelines. Agency procedures prepared pursuant to § 1500.3(a) of these guidelines shall implement these public information requirements and shall include arrangements for availability of environmental statements and comments at the head and appropriate regional offices of the responsible agency and at appropriate State and areawide clearinghouses unless the Governor of the State involved designates to the Council some other point for receipt of this information. Notice of such designation of an alternate point for receipt of this information will be included in the Office of Management and Budget listing of clearinghouses referred to in § 1500.9(c).

(e) Where emergency circumstances make it necessary to take an action with significant environmental impact without observing the provisions of these guidelines concerning minimum periods for agency review and advance availability of environmental statements, the Federal agency proposing to take the action should consult with the Council about alternative arrangements. Similarly where there are overriding considerations of expense to the Government or impaired program effectiveness, the responsible agency should consult with the Council concerning appropriate modifications of the minimum periods.

(f) In order to assist the Council in fulfilling its responsibilities under the Act and under Executive Order 11514, all agencies shall (as required by section 102(2)(H) of the Act and section 3(i) of Executive Order 11514) be responsive to requests by the Council for reports and other information dealing with issues arising in connection with the implementation of the Act. In particular, agencies shall be responsive to a request by the Council for the preparation and circulation of an environmental statement, unless the agency determines that such a statement is not required, in which case the agency shall prepare an environmental assessment and a publicly available record briefly setting forth the reasons for its determination. In no case, however, shall the Council's silence or failure to comment or request preparation, modification, or recirculation of an environmental statement or to take other action with respect to an environmental statement be construed as bearing in any way on the question of the legal requirement for or the adequacy of such statement under the Act.

§ 1500.12   Legislative actions.

(a) The Council and the Office of Management and Budget will cooperate in giving guidance as needed to assist agencies in identifying legislative items believed to have environmental significance. Agencies should prepare impact statements prior to submission of their legislative proposals to the Office of Management and Budget. In this regard, agencies should identify types of repetitive legislation requiring environmental impact statements (such as certain types of bills affecting transportation policy or annual construction authorizations).

(b) With respect to recommendations or reports on proposals for legislation to which section 102(2)(C) applies, the final text of the environmental statement and comments thereon should be available to the Congress and to the public for consideration in connection with the proposed legislation or report. In cases where the scheduling of congressional hearings on recommendations or reports on proposals for legislation which the Federal agency has forwarded to the Congress does not allow adequate time for the completion of a final text of an environmental statement (together with comments), a draft environmental statement may be furnished to the Congress and made available to the public pending transmittal of the comments as received and the final text.

§ 1500.13 Application of section 102(2)(C) procedure to existing projects and programs.

Agencies have an obligation to reassess ongoing projects and programs in order to avoid or minimize adverse environmental effects. The section 102(2)(C) procedure shall be applied to further major Federal actions having a significant effect on the environment even though they arise from projects or programs initiated prior to enactment of the Act on January 1, 1970. While the status of the work and degree of completion may be considered in determining whether to proceed with the project, it is essential that the environmental impacts of proceeding are reassessed pursuant to the Act's policies and procedures and, if the project or program is continued, that further incremental major actions be shaped so as to enhance and restore environmental quality as well as to avoid or minimize adverse environmental consequences. It is also important in further action that account be taken of environmental consequences not fully evaluated at the outset of the project or program.

§ 1500.14 Supplementary guidelines; evaluation of procedures.

(a) The Council after examining environmental statements and agency procedures with respect to such statements will issue such supplements to these guidelines as are necessary.

(b) Agencies will continue to assess their experience in the implementation of the section 102(2)(C) provisions of the Act and in conforming with these guidelines and report thereon to the Council by June 30, 1974. Such reports should include an identification of the problem areas and suggestions for revision or clarification of these guidelines to achieve effective coordination of views on environmental aspects (and alternatives, where appropriate) of proposed actions without imposing unproductive administrative procedures. Such reports shall also indicate what progress the agency has made in developing substantive criteria and guidance for making environmental assessments as required by § 1500.6(c) of this directive and by section 102(2)(B) of the Act.

Effective date. The revisions of these guidelines shall apply to all draft and final impact statements filed with the Council after January 28, 1973.

RUSSELL E. TRAIN,
Chairman.

APPENDIX I—SUMMARY TO ACCOMPANY DRAFT
AND FINAL STATEMENTS

(Check one) ( ) Draft. ( ) Final Environmental Statement.

Name of responsible Federal agency (with name of operating division where appropriate). Name, address, and telephone number of individual at the agency who can be contacted for additional information about the proposed action or the statement.

1. Name of action (Check one) ( ) Administrative Action. ( ) Legislative Action.

2. Brief description of action and its purpose. Indicate what States (and counties) particularly affected, and what other proposed Federal actions in the area, if any, are discussed in the statement.

3. Summary of environmental impacts and adverse environmental effects.

4. Summary of major alternatives considered.

5. (For draft statements) List all Federal, State, and local agencies and other parties from which comments have been requested. (For final statements) List all Federal, State, and local agencies and other parties from which written comments have been received.

6. Date draft statement (and final environmental statement, if one has been issued) made available to the Council and the public.

APPENDIX II—AREAS OF ENVIRONMENTAL IMPACT
AND FEDERAL AGENCIES AND FEDERAL STATE AGENCIES[1]
WITH JURISDICTION BY LAW OR SPECIAL EXPERTISE
TO COMMENT THEREON[2]

AIR

Air Quality

Department of Agriculture—
Forest Service (effects on vegetation)
Atomic Energy Commission (radioactive substances)
Department of Health, Education, and Welfare
Environmental Protection Agency
Department of the Interior—
Bureau of Mines (fossil and gaseous fuel combustion)
Bureau of Sport Fisheries and Wildlife (effect on wildlife)
Bureau of Outdoor Recreation (effects on recreation)
Bureau of Land Management (public lands)
Bureau of Indian Affairs (Indian lands)
National Aeronautics and Space Administration (remote sensing, aircraft
emissions)
Department of Transportation—
Assistant Secretary for Systems Development and Technology (auto
emissions)
Coast Guard (vessel emissions)
Federal Aviation Administration (aircraft emissions)

Weather Modification

Department of Agriculture—
Forest Service

---

[1] River Basin Commissions (Delaware, Great Lakes, Missouri, New
England, Ohio, Pacific Northwest, Souris-Red-Rainy, Susquehanna,
Upper Mississippi) and similar Federal-State agencies should be consulted on actions affecting the environment of their specific geographic
jurisdictions.

[2] In all cases where a proposed action will have significant international
environmental effects, the Department of State should be consulted, and
should be sent a copy of any draft and final impact statement which
covers such action.

Department of Commerce—
National Oceanic and Atmospheric Administration
Department of Defense—
Department of the Air Force
Department of the Interior
Bureau of Reclamation

WATER RESOURCES COUNCIL

WATER

Water Quality

Department of Agriculture—
Soil Conservation Service
Forest Service
Atomic Energy Commission (radioactive substances)
Department of the Interior—
Bureau of Reclamation
Bureau of Land Management (public lands)
Bureau of Indian Affairs (Indian lands)
Bureau of Sports Fisheries and Wildlife
Bureau of Outdoor Recreation
Geological Survey
Office of Saline Water
Environmental Protection Agency
Department of Health, Education, and Welfare
Department of Defense—
Army Corps of Engineers
Department of the Navy (ship pollution control)
National Aeronautics and Space Administration (remote sensing)
Department of Transportation—
Coast Guard (oil spills, ship sanitation)
Department of Commerce—
National Oceanic and Atmospheric Administration
Water Resources Council
River Basin Commissions (as geographically appropriate)

Marine Pollution, Commercial Fishery Conservation,
and Shellfish Sanitation

Department of Commerce—
National Oceanic and Atmospheric Administration

Department of Defense—
  Army Corps of Engineers
  Office of the Oceanographer of the Navy
Department of Health, Education, and Welfare
Department of the Interior—
  Bureau of Sport Fisheries and Wildlife
  Bureau of Outdoor Recreation
  Bureau of Land Management (outer continental shelf)
  Geological Survey (outer continental shelf)
Department of Transportation—
  Coast Guard
Environmental Protection Agency
National Aeronautics and Space Administration (remote sensing)
Water Resources Council
River Basin Commissions (as geographically appropriate)

Waterway Regulation and Stream Modification

Department of Agriculture—
  Soil Conservation Service
Department of Defense—
  Army Corps of Engineers
Department of the Interior—
  Bureau of Reclamation
  Bureau of Sport Fisheries and Wildlife
  Bureau of Outdoor Recreation
  Geological Survey
Department of Transportation—
  Coast Guard
Environmental Protection Agency
National Aeronautics and Space Administration (remote sensing)
Water Resources Council
River Basin Commissions (as geographically appropriate)

FISH AND WILDLIFE

Department of Agriculture—
  Forest Service
  Soil Conservation Service
Department of Commerce—
  National Oceanic and Atmospheric Administration (marine species)
Department of the Interior—
  Bureau of Sport Fisheries and Wildlife

Bureau of Land Management
Bureau of Outdoor Recreation
Environmental Protection Agency

SOLID WASTE

Atomic Energy Commission (radioactive waste)
Department of Defense—
Army Corps of Engineers
Department of Health, Education, and Welfare
Department of the Interior—
Bureau of Mines (mineral waste, mine acid waste, municipal solid waste, recycling)
Bureau of Land Management (public lands)
Bureau of Indian Affairs (Indian lands)
Geological Survey (geologic and hydrologic effects)
Office of Saline Water (demineralization)
Department of Transportation—
Coast Guard (ship sanitation)
Environmental Protection Agency
River Basin Commissions (as geographically appropriate)
Water Resources Council

NOISE

Department of Commerce—
National Bureau of Standards
Department of Health, Education, and Welfare
Department of Housing and Urban Development (land use and building materials aspects)
Department of Labor—
Occupational Safety and Health Administration
Department of Transportation—
Assistant Secretary for Systems Development and Technology
Federal Aviation Administration, Office of Noise Abatement
Environmental Protection Agency
National Aeronautics and Space Administration

RADIATION

Atomic Energy Commission
Department of Commerce—
National Bureau of Standards
Department of Health, Education, and Welfare

Department of the Interior—
  Bureau of Mines (uranium mines)
  Mining Enforcement and Safety Administration (uranium mines)
Environmental Protection Agency

HAZARDOUS SUBSTANCES

Toxic Materials

Atomic Energy Commission (radioactive substances)
Department of Agriculture—
  Agricultural Research Service
  Consumer and Marketing Service
Department of Commerce—
  National Oceanic and Atmospheric Administration
Department of Defense
Department of Health, Education, and Welfare
Environmental Protection Agency

Food Additives and Contamination of Foodstuffs

Department of Agriculture—
  Consumer and Marketing Service (meat and poultry products)
Department of Health, Education, and Welfare
Environmental Protection Agency

Pesticides

Department of Agriculture—
  Agricultural Research Service (biological controls, food and fiber production)
  Consumer and Marketing Service
  Forest Service
Department of Commerce—
  National Oceanic and Atmospheric Administration
Department of Health, Education, and Welfare
Department of the Interior—
  Bureau of Sport Fisheries and Wildlife (fish and wildlife effects)
  Bureau of Land Management (public lands)
  Bureau of Indian Affairs (Indian lands)
  Bureau of Reclamation (irrigated lands)
Environmental Protection Agency

Transportation and Handling of Hazardous Materials

Atomic Energy Commission (radioactive substances)
Department of Commerce—
  Maritime Administration
  National Oceanic and Atmospheric Administration (effects on marine
    life and the coastal zone)
Department of Defense—
  Armed Services Explosive Safety Board
  Army Corps of Engineers (navigable waterways)
Department of Transportation—
  Federal Highways Administration, Bureau of Motor Carrier Safety
  Coast Guard
  Federal Railroad Administration
  Federal Aviation Administration
  Assistant Secretary for Systems Development and Technology
  Office of Hazardous Materials
  Office of Pipeline Safety
Environment Protection Agency

ENERGY SUPPLY AND NATURAL RESOURCES DEVELOPMENT

Electric Energy Development, Generation, and Transmission, and Use

Atomic Energy Commission (nuclear)
Department of Agriculture—
  Rural Electrification Administration (rural areas)
Department of Defense—
  Army Corps of Engineers (hydro)
Department of Health, Education, and Welfare (radiation effects)
Department of Housing and Urban Development (urban areas)
Department of the Interior—
  Bureau of Indian Affairs (Indian lands)
  Bureau of Land Management (public lands)
  Bureau of Reclamation
  Power Marketing Administrations
  Geological Survey
  Bureau of Sport Fisheries and Wildlife
  Bureau of Outdoor Recreation
  National Park Service
Environmental Protection Agency
Federal Power Commission (hydro, transmission, and supply)

River Basin Commissions (as geographically appropriate)
Tennessee Valley Authority
Water Resources Council

Petroleum Development, Extraction, Refining, Transport, and Use

Department of the Interior—
  Office of Oil and Gas
  Bureau of Mines
  Geological Survey
  Bureau of Land Management (public lands and outer continental shelf)
  Bureau of Indian Affairs (Indian lands)
  Bureau of Sport Fisheries and Wildlife (effects on fish and wildlife)
Bureau of Outdoor Recreation
  National Park Service
Department of Transportation (Transport and Pipeline Safety)
Environmental Protection Agency
Interstate Commerce Commission

Natural Gas Development, Production, Transmission, and Use

Department of Housing and Urban Development (urban areas)
Department of the Interior—
  Office of Oil and Gas
  Geological Survey
  Bureau of Mines
  Bureau of Land Management (public lands)
  Bureau of Indian Affairs (Indian lands)
  Bureau of Sport Fisheries and Wildlife
  Bureau of Outdoor Recreation
  National Park Service
Department of Transportation (transport and safety)
Environmental Protection Agency
Federal Power Commission (production, transmission, and supply)
Interstate Commerce Commission

*Coal and Minerals Development, Mining, Conversion,
Processing, Transport, and Use*

Appalachian Regional Commission
Department of Agriculture—
  Forest Service
Department of Commerce

Department of the Interior—
 Office of Coal Research
 Mining Enforcement and Safety Administration
 Bureau of Mines
 Geological Survey
 Bureau of Indian Affairs (Indian lands)
 Bureau of Land Management (public lands)
 Bureau of Sport Fisheries and Wildlife
 Bureau of Outdoor Recreation
 National Park Service
Department of Labor—
 Occupational Safety and Health Administration
Department of Transportation
Environmental Protection Agency
Interstate Commerce Commission
Tennessee Valley Authority

*Renewable Resource Development, Production, Management, Harvest,
Transport, and Use*

Department of Agriculture—
 Forest Service
 Soil Conservation Service
Department of Commerce
Department of Housing and Urban Development (building materials)
Department of the Interior—
 Geological Survey
 Bureau of Land Management (public lands)
 Bureau of Indian Affairs (Indian lands)
 Bureau of Sport Fisheries and Wildlife
 Bureau of Outdoor Recreation
 National Park Service
Department of Transportation
Environmental Protection Agency
Interstate Commerce Commission (freight rates)

Energy and Natural Resources Conservation

Department of Agriculture—
 Forest Service
 Soil Conservation Service
Department of Commerce—
 National Bureau of Standards (energy efficiency)

Department of Housing and Urban Development—
  Federal Housing Administration (housing standards)
Department of the Interior—
  Office of Energy Conservation
  Bureau of Mines
  Bureau of Reclamation
  Geological Survey
  Power Marketing Administration
Department of Transportation
Environmental Protection Agency
Federal Power Commission
General Services Administration (design and operation of buildings)
Tennessee Valley Authority

LAND USE AND MANAGEMENT

Land Use Changes, Planning and Regulation of Land Development

Department of Agriculture—
  Forest Service (forest lands)
  Agricultural Research Service (agricultural lands)
Department of Housing and Urban Development
Department of the Interior—
  Office of Land Use and Water Planning
  Bureau of Land Management (public lands)
  Bureau of Indian Affairs (Indian lands)
  Bureau of Sport Fisheries and Wildlife (wildlife refuges)
  Bureau of Outdoor Recreation (recreation lands)
  National Park Service (NPS units)
Department of Transportation
Environmental Protection Agency (pollution effects)
National Aeronautics and Space Administration (remote sensing)
River Basin Commissions (as geographically appropriate).

Public Land Management

Department of Agriculture—
  Forest Service (forests)
Department of Defense
Department of the Interior—
  Bureau of Land Management
  Bureau of Indian Affairs (Indian lands)

Bureau of Sport Fisheries and Wildlife (wildlife refuges)
Bureau of Outdoor Recreation (recreation lands)
National Park Service (NPS units)
Federal Power Commission (project lands)
General Services Administration
National Aeronautics and Space Administration (remote sensing)
Tennessee Valley Authority (project lands)

PROTECTION OF ENVIRONMENTALLY CRITICAL AREAS—FLOODPLAINS, WETLANDS, BEACHES AND DUNES, UNSTABLE SOILS, STEEP SLOPES, AQUIFER RECHARGE AREAS, ETC.

Department of Agriculture—
  Agricultural Stabilization and Conservation Service
  Soil Conservation Service
  Forest Service
Department of Commerce—
  National Oceanic and Atmospheric Administration (coastal areas)
Department of Defense—
  Army Corps of Engineers
Department of Housing and Urban Development (urban and floodplain
    areas)
Department of the Interior—
  Office of Land Use and Water Planning
  Bureau of Outdoor Recreation
  Bureau of Reclamation
  Bureau of Sport Fisheries and Wildlife
  Bureau of Land Management
  Geological Survey
Environmental Protection Agency (pollution effects)
National Aeronautics and Space Administration (remote sensing)
River Basin Commissions (as geographically appropriate)
Water Resources Council

LAND USE IN COASTAL AREAS

Department of Agriculture—
  Forest Service
  Soil Conservation Service (soil stability, hydrology)
Department of Commerce—
  National Oceanic and Atmospheric Administration (impact on marine
    life and coastal zone management)

Department of Defense—
  Army Corps of Engineers (beaches, dredge and fill permits, Refuse Act
    permits)
Department of Housing and Urban Development (urban areas)
Department of the Interior—
  Office of Land Use and Water Planning
  Bureau of Sport Fisheries and Wildlife
  National Park Service
  Geological Survey
  Bureau of Outdoor Recreation
  Bureau of Land Management (public lands)
Department of Transportation—
  Coast Guard (bridges, navigation)
Environmental Protection Agency (pollution effects)
National Aeronautics and Space Administration (remote sensing)

REDEVELOPMENT AND CONSTRUCTION IN BUILT-UP AREAS

Department of Commerce—
  Economic Development Administration (designated areas)
Department of Housing and Urban Development
Department of the Interior—
  Office of Land Use and Water Planning
Department of Transportation
Environmental Protection Agency
General Services Administration
Office of Economic Opportunity

DENSITY AND CONGESTION MITIGATION

Department of Health, Education, and Welfare
Department of Housing and Urban Development
Department of the Interior—
  Office of Land Use and Water Planning
  Bureau of Outdoor Recreation
Department of Transportation
Environmental Protection Agency

NEIGHBORHOOD CHARACTER AND CONTINUITY

Department of Health, Education, and Welfare
Department of Housing and Urban Development

National Endowment for the Arts
Office of Economic Opportunity

IMPACTS ON LOW-INCOME POPULATIONS

Department of Commerce—
  Economic Development Administration (designated areas)
Department of Health, Education, and Welfare
Department of Housing and Urban Development
Office of Economic Opportunity

HISTORIC, ARCHITECTURAL, AND ARCHEOLOGICAL PRESERVATION

Advisory Council on Historic Preservation
Department of Housing and Urban Development
Department of the Interior—
  National Park Service
  Bureau of Land Management (public lands)
  Bureau of Indian Affairs (Indian lands)
General Services Administration
National Endowment for the Arts

SOIL AND PLANT CONSERVATION AND HYDROLOGY

Department of Agriculture—
  Soil Conservation Service
  Agricultural Service
  Forest Service
Department of Commerce—
  National Oceanic and Atmospheric Administration
Department of Defense—
  Army Corps of Engineers (dredging, aquatic plants)
Department of Health, Education, and Welfare
Department of the Interior—
  Bureau of Land Management
  Bureau of Sport Fisheries and Wildlife
  Geological Survey
  Bureau of Reclamation
Environmental Protection Agency
National Aeronautics and Space Administration (remote sensing)
River Basin Commissions (as geographically appropriate)
Water Resources Council

OUTDOOR RECREATION

Department of Agriculture—
  Forest Service
  Soil Conservation Service
Department of Defense—
  Army Corps of Engineers
Department of Housing and Urban Development (urban areas)
Department of the Interior—
  Bureau of Land Management
  National Park Service
  Bureau of Outdoor Recreation
  Bureau of Sport Fisheries and Wildlife
  Bureau of Indian Affairs
Environmental Protection Agency
National Aeronautics and Space Administration (remote sensing)
River Basin Commissions (as geographically appropriate)
Water Resources Council

APPENDIX III—OFFICES WITHIN FEDERAL AGENCIES AND FEDERAL-
  STATE AGENCIES FOR INFORMATION REGARDING THE AGENCIES'
  NEPA ACTIVITIES AND FOR RECEIVING OTHER AGENCIES' IMPACT
  STATEMENTS FOR WHICH COMMENTS ARE REQUESTED

ADVISORY COUNCIL ON HISTORIC PRESERVATION

Office of Architectural and Environmental Preservation, Advisory
  Council on Historic Preservation, Suite 430, 1522 K Street, N.W.,
  Washington, D.C. 20005 254-3974.

DEPARTMENT OF AGRICULTURE[1]

Office of the Secretary, Attn: Coordinator Environmental Quality Activities. U.S. Department of Agriculture, Washington, D.C. 20250
  447-3965

APPALACHIAN REGIONAL COMMISSION

Office of the Alternate Federal Co-Chairman, Appalachian Regional
  Commission, 1666 Connecticut Avenue, N.W., Washington, D.C.
  20235 967-4103

---

[1] Requests for comments or information from individual units of the
Department of Agriculture, e.g., Soil Conservation Service, Forest
Service, etc. should be sent to the Office of the Secretary, Department of
Agriculture, at the address given above.

DEPARTMENT OF THE ARMY (CORPS OF ENGINEERS)

Executive Director of Civil Works, Office of the Chief of Engineers, U.S. Army Corps of Engineers, Washington, D.C. 20314 693-7168

ATOMIC ENERGY COMMISSION

For nonregulatory matters: Office of Assistant General Manager for Biomedical and Environmental Research and Safety Programs, Atomic Energy Commission, Washington, D.C. 20545 973-3208
For regulatory matters: Office of the Assistant Director for Environmental Projects, Atomic Energy Commission, Washington, D.C. 20545 973-7531

DEPARTMENT OF COMMERCE

Office of the Deputy Assistant Secretary for Environmental Affairs, U.S. Department of Commerce, Washington, D.C. 20230 967-4335

DEPARTMENT OF DEFENSE

Office of the Assistant Secretary for Defense (Health and Environment). U.S. Department of Defense, Room 3E172, The Pentagon, Washington, D.C. 20301 697-2111

DELAWARE RIVER BASIN COMMISSION

Office of the Secretary, Delaware River Basin Commission, Post Office Box 360, Trenton, N.J. 08603
(609)883-9500

ENVIRONMENTAL PROTECTION AGENCY[2]

Director, Office of Federal Activities, Environmental Protection Agency, 401 M Street, S.W., Washington, D.C. 20460 755-0777

---

[2] Contact the Office of Federal Activities for environmental statements concerning legislation, regulations, national program proposals or other major policy issues.

For all other EPA consultation, contact the Regional Administrator in whose area the proposed action (e.g., highway or water resource construction projects) will take place. The Regional Administrators will coordinate the EPA review. Addresses of the Regional Administrators, and the areas covered by their regions are as follows:

Footnote 2 Continued

Regional Administrator, I,
  U.S. Environmental Protection Agency
  Room 2303, John F. Kennedy
  Federal Bldg., Boston, Mass. 02203,
  (617)233-7210

Connecticut, Maine,
  Massachusetts,
  New Hampshire,
  Rhode Island, Vermont

Regional Administrator, II,
  U.S. Environmental Protection Agency
  Room 908, 26 Federal Plaza
  New York, New York 10007
  (212)264-2525

New Jersey,
New York,
Puerto Rico,
Virgin Islands

Regional Administrator III.
  U.S. Environmental Protection Agency
  Curtis Bldg., 6th & Walnut Sts.
  Philadelphia, Pa. 19106
  (215) 597-9801

Delaware, Maryland,
  Pennsylvania, Virginia,
  West Virginia,
  District of Columbia

Regional Administrator IV,
  U.S. Environmental Protection Agency
  1421 Peachtree Street N.E.
  Atlanta, Ga. 30309
  (404) 526-5727

Alabama, Florida, Georgia,
  Kentucky, Mississippi,
  North Carolina, South
  Carolina, Tennessee

Regional Administrator V,
  U.S. Environmental Protection Agency
  1 N. Wacker Drive
  Chicago, Illinois 60606
  (312) 353-5250

Illinois, Indiana, Michigan,
  Minnesota, Ohio,
  Wisconsin

Regional Administrator VI,
  U.S. Environmental Protection Agency
  1600 Patterson Street
  Suite 1100
  Dallas, Texas 75201
  (214) 749-1962

Arkansas, Louisiana, New
  Mexico, Texas,
  Oklahoma

Regional Administrator VII,
  U.S. Environmental Protection Agency
  1735 Baltimore Avenue
  Kansas City, Missouri 64108
  (816) 374-5493

Iowa, Kansas, Missouri,
  Nebraska

Regional Administrator VIII,
  U.S. Environmental Protection Agency
  Suite 900, Lincoln Tower
  1860 Lincoln Street
  Denver, Colorado 80203
  (303) 837-3895

Colorado, Montana, North
  Dakota, South Dakota,
  Utah, Wyoming

FEDERAL POWER COMMISSION

Commission's Advisor on Environmental Quality, Federal Power Commission, 825 N. Capitol Street, N.E., Washington, D.C. 20426 386-6084

GENERAL SERVICES ADMINISTRATION

Office of Environmental Affairs, Office of the Deputy Administrator for Special Projects, General Services Administration, Washington, D.C. 20405 343-4161

GREAT LAKES BASIN COMMISSION

Office of the Chairman, Great Lakes Basin Commission, 3475 Plymouth Road, P.O. Box 999, Ann Arbor, Michigan 48105 (313)769-7431

DEPARTMENT OF HEALTH, EDUCATION AND WELFARE[3]

Office of Environmental Affairs, Office of the Assistant Secretary for Administration and Management, Department of Health, Education and Welfare, Washington, D.C. 20202 963-4456

---

[3] Contact the Office of Environmental Affairs for information on HEW's environmental statements concerning legislation, regulations, national program proposals or other major policy issues, and for all requests for HEW comment on impact statements of other agencies.

For information with respect to HEW actions occurring within the jurisdiction of the Departments' Regional Directors, contact the appropriate Regional Environmental Officer:

Footnote 2 Continued

Regional Administrator IX,
    U.S. Environmental Protection Agency
    100 California Street
    San Francisco, California 94111
    (415) 556-2320
Arizona, California, Hawaii, Nevada, American Samoa, Guam, Trust Territories of Pacific Islands, Wake Island

Regional Administrator X,
    U.S. Environmental Protection Agency
    1200 Sixth Avenue
    Seattle, Washington 98101
    (206) 442-1220
Alaska, Idaho, Oregon, Washington

Footnote 3 Continued
Region I:
    Regional Environmental Officer
    U.S. Department of Health, Education and Welfare
    Room 2007B
    John F. Kennedy Center
    Boston, Massachusetts 02203 (617)223-6837
Region II:
    Regional Environmental Officer
    U.S. Department of Health, Education and Welfare
    Federal Building
    26 Federal Plaza
    New York, New York 10007 (212) 264-1308
Region III:
    Regional Environmental Officer
    U.S. Department of Health, Education and Welfare
    P.O. Box 13716
    Philadelphia, Pennsylvania 19101 (215) 597-6498
Region IV:
    Regional Environmental Officer
    U.S. Department of Health, Education and Welfare
    Room 404
    50 Seventh Street, N.E.
    Atlanta, Georgia 30323 (404) 526-5817
Region V:
    Regional Environmental Officer
    U.S. Department of Health, Education and Welfare
    Room 712, New Post Office Building
    433 West Van Buren Street
    Chicago, Illinois 60607 (312) 353-1644
Region VI:
    Regional Environmental Officer
    U.S. Department of Health, Education and Welfare
    1114 Commerce Street
    Dallas, Texas 75202 (214) 749-2236
Region VII:
    Regional Environmental Officer
    U.S. Department of Health, Education and Welfare
    601 East 12th Street
    Kansas City, Missouri 64106 (816) 374-3584

Director, Office of Community and Environmental Standards, Department of Housing and Urban Development, Room 7206, Washington, D.C. 20410 755-5980

---

[4] Contact the Director with regard to environmental impacts of legislation, policy statements, program regulations and procedures, and precedent-making project decisions. For all other HUD consultation, contact the HUD Regional Administrator in whose jurisdiction the project lies, as follows:
Regional Administrator I,
Environmental Clearance Officer
U.S. Department of Housing and Urban Development
Room 405, John F. Kennedy Federal Building
Boston, Mass. 02203 (617) 223-4066

Footnote 3 Continued
Region VIII:
Regional Environmental Officer
U.S. Department of Health, Education and Welfare
9017 Federal Building
19th and Stout Streets
Denver, Colorado 80202 (303) 837-4178
Region IX:
Regional Environmental Officer
U.S. Department of Health, Education and Welfare
50 Fulton Street
San Francisco, California 94102 (415) 556-1970
Region X:
Regional Environmental Officer
U.S. Department of Health, Education and Welfare
Arcade Plaza Building
1321 Second Street
Seattle, Washington 98101 (206) 442-0490

Footnote 4 Continued
Regional Administrator II,
   Environmental Clearance Officer
   U.S. Department of Housing and Urban Development
   26 Federal Plaza
   New York, New York 10007 (212) 264-8068
Regional Administrator III,
   Environmental Clearance Officer
   U.S. Department of Housing and Urban Development
   Curtis Building, Sixth and Walnut Street
   Philadelphia, Pennsylvania 19106 (215) 597-2560
Regional Administrator IV,
   Environmental Clearance Officer
   U.S. Department of Housing and Urban Development
   Peachtree-Seventh Building
   Atlanta, Georgia 30323 (404) 526-5585
Regional Administrator V,
   Environmental Clearance Officer
   U.S. Department of Housing and Urban Development
   360 North Michigan Avenue
   Chicago, Illinois 60601 (312) 353-5680
Regional Administrator VI,
   Environmental Clearance Officer
   U.S. Department of Housing and Urban Development
   Federal Office Building, 819 Taylor Street
   Fort Worth, Texas 76102 (817) 334-2867
Regional Administrator VII,
   Environmental Clearance Officer
   U.S. Department of Housing and Urban Development
   911 Walnut Street
   Kansas City, Missouri 64106 (816) 374-2661
Regional Administrator VIII,
   Environmental Clearance Officer
   U.S. Department of Housing and Urban Development
   Samsonite Building, 1051 South Broadway
   Denver, Colorado 80209 (303) 837-4061
Regional Administrator IX,
   Environmental Clearance Officer
   U.S. Department of Housing and Urban Development
   450 Golden Gate Avenue, Post Office Box 36003
   San Francisco, California 94102 (415) 556-4752

DEPARTMENT OF THE INTERIOR[5]

Director, Office of Environmental Project Review, Department of the Interior, Interior Building, Washington, D.C. 20240 343-3891

INTERSTATE COMMERCE COMMISSION

Office of Proceedings, Interstate Commerce Commission, Washington, D.C. 20423
343-6167

DEPARTMENT OF LABOR

Assistant Secretary for Occupational Safety and Health, Department of Labor, Washington, D.C. 20210
961-3405

MISSOURI RIVER BASINS COMMISSION

Office of the Chairman, Missouri River Basins Commission, 10050 Regency Circle, Omaha, Nebraska 68114
(402) 397-5714

NATIONAL AERONAUTICS AND SPACE ADMINISTRATION

Office of the Comptroller, National Aeronautics and Space Administration, Washington, D.C. 20546
755-8440

NATIONAL CAPITAL PLANNING COMMISSION

Office of Environmental Affairs, Office of the Executive Director, National Capital Planning Commission, Washington, D.C. 20576
382-7200

---

[5] Requests for comments or information from individual units of the Department of the Interior should be sent to the Office of Environmental Project Review at the address given above.

Footnote 4 Continued
Regional Administrator X,
    Environmental Clearance Officer
    U.S. Department of Housing and Urban Development
    Room 226, Arcade Plaza Building
    Seattle, Washington 28101 (206) 583-5415

NATIONAL ENDOWMENT FOR THE ARTS

Office of Architecture and Environmental Arts Program, National Endowment for the Arts, Washington, D.C. 20506
382-5765

NEW ENGLAND RIVER BASINS COMMISSION

Office of the Chairman, New England River Basins Commission, 55 Court Street, Boston, Mass. 02108
(617)223-6244

OFFICE OF ECONOMIC OPPORTUNITY

Office of the Director, Office of Economic Opportunity, 1200 19th Street N.W., Washington, D.C. 20506
254-6000

OHIO RIVER BASIN COMMISSION

Office of the Chairman, Ohio River Basin Commission, 36 East 4th Street, Suite 208-20, Cincinnati, Ohio 45202
(513) 684-3831

PACIFIC NORTHWEST RIVER BASINS COMMISSION

Office of the Chairman, Pacific Northwest River Basins Commission, 1 Columbia River, Vancouver, Washington 98660
(206) 695-3606

SOURIS-RED-RAINY RIVER BASINS COMMISSION

Office of the Chairman, Souris-Red-Rainy River Basins Commission, Suite 6, Professional Building, Holiday Mall, Moorhead, Minnesota 56560
(701) 237-5227

DEPARTMENT OF STATE

Office of the Special Assistant to the Secretary for Environmental Affairs, Department of State, Washington, D.C. 20520
632-7964

SUSQUEHANNA RIVER BASIN COMMISSION

Office of the Executive Director, Susquehanna River Basin Commission, 5012 Lenker Street, Mechanicsburg, Pa. 17055
(717) 737-0501

TENNESSEE VALLEY AUTHORITY

Office of the Director of Environmental Research and Development, Tennessee Valley Authority, 720 Edney Building, Chattanooga, Tennessee 37401 (615) 755-2002

DEPARTMENT OF TRANSPORTATION[6]

Director, Office of Environmental Quality, Office of the Assistant Secretary for Environment, Safety, and Consumer Affairs, Department of Transportation, Washington, D.C. 20590 426-4357

---

[6] Contact the Office of Environmental Quality, Department of Transportation, for information on DOT's environmental statements concerning legislation, regulations, national program proposals, or other major policy issues.

For information regarding the Department of Transportation's other environmental statements, contact the national office for the appropriate administration:

U.S. Coast Guard

Office of Marine Environment and Systems, U.S. Coast Guard, 400 7th Street, S.W., Washington, D.C. 20590, 426-2007

Federal Aviation Administration

Office of Environmental Quality, Federal Aviation Administration, 800 Independence Avenue, S.W., Washington, D.C. 20591, 426-8406

Federal Highway Administration

Office of Environmental Policy, Federal Highway Administration, 400 7th Street, S.W., Washington, D.C. 20590, 426-0351

Federal Railroad Administration

Office of Policy and Plans, Federal Railroad Administration, 400 7th Street, S.W., Washington, D.C. 20590, 426-1567

Urban Mass Transportation Administration

Office of Program Operations, Urban Mass Transportation Administration, 400 7th Street, S.W., Washington, D.C. 20590, 426-4020

For other administration's not listed above, contact the Office of Environmental Quality, Department of Transportation, at the address given above.

Footnote 6 Continued

For comments on other agencies' environmental statements, contact the appropriate administration's regional office. If more than one administration within the Department of Transportation is to be requested to comment, contact the Secretarial Representative in the appropriate Regional Office for coordination of the Department's comments:

Secretarial Representative

Region I Secretarial Representative, U.S. Department of Transportation, Transportation Systems Center, 55 Broadway, Cambridge, Massachusetts 02142 (617) 494-2709

Region II Secretarial Representative, U.S. Department of Transportation, 26 Federal Plaza, Room 1811, New York, New York 10007 (212) 264-2672

Region III Secretarial Representative, U.S. Department of Transportation, Mall Building, Suite 1214, 325 Chestnut Street, Philadelphia, Pennsylvania 19106 (215) 597-0407

Region IV Secretarial Representative, U.S. Department of Transportation, Suite 515, 1720 Peachtree Rd., N.W. Atlanta, Georgia 30309 (404) 526-3738

Region V Secretarial Representative, U.S. Department of Transportation, 17th Floor, 300 S. Wacker Drive, Chicago, Illinois 60606 (312) 353-4000

Region VI Secretarial Representative, U.S. Department of Transportation, 9-C-18 Federal Center, 1100 Commerce Street, Dallas, Texas 75202 (214) 749-1851

Region VII Secretarial Representative, U.S. Department of Transportation, 601 E. 12th Street, Room 634, Kansas City, Missouri 64106 (816) 374-2761

Region VIII Secretarial Representative, U.S. Department of Transportation, Prudential Plaza, Suite 1822, 1050 17th Street, Denver, Colorado 80225 (303) 837-3242

Region IX Secretarial Representative, U.S. Department of Transportation, 450 Golden Gate Avenue, Box 36133, San Francisco, California 94102 (415) 556-5961

Region X Secretarial Representative, U.S. Department of Transportation, 1321 Second Avenue, Room 507, Seattle, Washington 98101 (206) 442-0590

FEDERAL AVIATION ADMINISTRATION

New England Region, Office of the Regional Director, Federal Aviation Administration, 154 Middlesex Street, Burlington, Massachusetts 01803 (617) 272-2350

Footnote 6 Continued

Eastern Region, Office of the Regional Director, Federal Aviation Administration, Federal Building, JFK International Airport, Jamaica, New York 11430 (212) 995-3333

Southern Region, Office of the Regional Director, Federal Aviation Administration, P.O. Box 20636, Atlanta, Georgia 30320 (404) 526-7222

Great Lakes Region, Office of the Regional Director, Federal Aviation Administration, 2300 East Devon, Des Plaines, Illinois 60018 (312) 694-4500

Southwest Region, Office of the Regional Director, Federal Aviation Administration, P.O. Box 1689, Fort Worth, Texas 76101 (817) 624-4911

Central Region, Office of the Regional Director, Federal Aviation Administration, 601 E. 12th Street, Kansas City, Missouri 64106 (816) 374-5626

Rocky Mountain Region, Office of the Regional Director, Federal Aviation Administration, Park Hill Station, P.O. Box 7213, Denver, Colorado 80207 (303) 837-3646

Western Region, Office of the Regional Director, Federal Aviation Administration, P.O. Box 92007, WorldWay Postal Center, Los Angeles, California 90009 (213) 536-6427

Northwest Region, Office of the Regional Director, Federal Aviation Administration, FAA Building, Boeing Field, Seattle, Washington 98108 (206) 767-2780

FEDERAL HIGHWAY ADMINISTRATION

Region 1, Regional Administrator, Federal Highway Administration, 4 Normanskill Boulevard, Delmar, New York 12054 (518) 472-6476

Region 2, Regional Administrator, Federal Highway Administration, Room 1621, George H. Fallon Federal Office Building, 31 Hopkins Plaza, Baltimore, Maryland 21201 (301) 962-2361

Region 3, Regional Administrator, Federal Highway Administration, Suite 200, 1720 Peachtree Road, N.W., Atlanta, Georgia 30309 (404) 526-5078

Region 4, Regional Administrator, Federal Highway Administration, Dixie Highway, Homewood, Illinois 60430 (312) 799-6300

Region 5, Regional Administrator, Federal Highway Administration, 819 Taylor Street, Fort Worth, Texas 76102 (817) 334-3232

Region 6, Regional Administrator, Federal Highway Administration, P.O. Box 7186, Country Club Station, Kansas City, Missouri 64113 (816) 361-7563

Region 7, Regional Administrator, Federal Highway Administration, Room 242, Building 40, Denver Federal Center, Denver, Colorado 80225

Footnote 6 Continued

Region 8, Regional Administrator, Federal Highway Administration, 450 Golden Gate Avenue, Box 36096, San Francisco, California 94102 (415) 556-3894

Region 9, Regional Administrator, Federal Highway Administration, Room 412, Mohawk Building, 222 S.W. Morrison Street, Portland, Oregon 97204 (503) 221-2065

URBAN MASS TRANSPORTATION ADMINISTRATION

Region I, Office of the UMTA Representative, Urban Mass Transportation Administration, Transportation Systems Center, Technology Building, Room 277, 55 Broadway, Boston, Massachusetts 02142 (617) 494-2055

Region II, Office of the UMTA Representative, Urban Mass Transportation Administration, 26 Federal Plaza, Suite 1809, New York, New York 10007 (212) 264-8162

Region III, Office of the UMTA Representative, Urban Mass Transportation Administration, Mall Building, Suite 1214, 325 Chestnut Street, Philadelphia, Pennsylvania 19106 (215) 597-0407

Region IV, Office of UMTA Representative, Urban Mass Transportation Administration, 1720 Peachtree Road, Northwest, Suite 501, Atlanta, Georgia 30309 (404) 526-3948

Region V, Office of the UMTA Representative, Urban Mass Transportation Administration, 300 South Wacker Drive, Suite 700, Chicago, Illinois 60606 (312) 353-6005

Region VI, Office of the UMTA Representative, Urban Mass Transportation Administration, Federal Center, Suite 9E24, 1100 Commerce Street, Dallas, Texas 75202 (214) 749-7322

Region VII, Office of the UMTA Representative, Urban Mass Transportation Administration, c/o FAA Management Systems Division, Room 1564D, 601 East 12th Street, Kansas City, Missouri 64106 (816) 374-5567

Region VIII, Office of the UMTA Representative, Urban Mass Transportation Administration, Prudential Plaza, Suite 1822, 1050 17th Street, Denver, Colorado 80202 (303) 837-3242

Region IX, Office of the UMTA Representative, Urban Mass Transportation Administration, 450 Golden Gate Avenue, Box 36125, San Francisco, California 94102 (415) 556-2884

Region X, Office of the UMTA Representative, Urban Mass Transportation Administration, 1321 Second Avenue, Suite 5079, Seattle, Washington 98101 (206) 442-0590

DEPARTMENT OF THE TREASURY

Office of Assistant Secretary for Administration, Department of the Treasury, Washington, D.C., 20220 964-5391

UPPER MISSISSIPPI RIVER BASIN COMMISSION

Office of the Chairman, Upper Mississippi River Basin Commission, Federal Office Building, Fort Snelling, Twin Cities, Minnesota 55111 (612) 725-4690

WATER RESOURCES COUNCIL

Office of the Associate Director, Water Resources Council, 2120 L Street, N.W., Suite 800, Washington, D.C. 20037 254-6442

APPENDIX IV—STATE AND LOCAL AGENCY REVIEW OF IMPACT STATEMENTS

1. OMB Circular No. A-95 through its system of clearinghouses provides a means for securing the views of State and local environmental agencies, which can assist in the preparation of impact statements. Under A-95, review of the proposed project in the case of federally assisted projects (Part I of A-95) generally takes place prior to the preparation of the impact statement. Therefore, comments on the environmental effects of the proposed project that are secured during this stage of the A-95 process represent inputs to the environmental impact statement.

2. In the case of direct Federal development (Part II of A-95), Federal agencies are required to consult with clearinghouses at the earliest practicable time in the planning of the project or activity. Where such consultation occurs prior to completion of the draft impact statement, comments relating to the environmental effects of the proposed action would also represent inputs to the environmental impact statement.

3. In either case, whatever comments are made on environmental effects of proposed Federal or federally assisted projects by clearinghouses, or by State and local environmental agencies through clearinghouses, in the course of the A-95 review should be attached to the draft impact statement when it is circulated for review. Copies of the statement should be sent to the agencies making such comments. Whether those agencies then elect to comment again on the basis of the draft impact statement is a matter to be left to the discretion of the commenting agency depending on its resources, the significance of the project, and the extent to which its earlier comments were considered in preparing the draft statement.

4. The clearinghouses may also be used, by mutual agreement, for securing reviews of the draft environmental impact statement. However, the Federal agency may wish to deal directly with appropriate State or local agencies in the review of impact statements because the clearinghouses may be unwilling or unable to handle this phase of the process. In some cases, the Governor may have designated a specific agency, other than the clearinghouse, for securing reviews of impact statements. In any case, the clearinghouses should be sent copies of the impact statement.

5. To aid clearinghouses in coordinating State and local comments, draft statements should include copies of State and local agency comments made earlier under the A-95 process and should indicate on the summary sheet those other agencies from which comments have been requested, as specified in Appendix I of the CEQ Guidelines.

[FR Doc.73-15783 Filed 7-31-73;8:45am]

Index

INDEX

361

A